ELECTRONIC COM
AND SYST

ELECTRONIC COMPONENTS AND SYSTEMS

W. H. Dennis, BSc (Lond.)

Butterworth Scientific
London Boston Sydney Wellington Durban Toronto

All rights reserved. No part of this publication may be reproduced or transmitted in any form or by any means, including photocopying and recording without the written permission of the copyright holder, application for which should be addressed to the publishers. Such written permission must also be obtained before any part of this publication is stored in a retrieval system of any nature.

This book is sold subject to the Standard Conditions of Sale of Net Books and may not be resold in the UK below the net price given by the Publishers in their current price list.

First published 1982
© Butterworth & Co (Publishers) Ltd 1982

British Library Cataloguing in Publication Data

Dennis, W.H.
  Electronic components and systems.
  1. Electronic circuits
  I. Title
  621.3815'3   TK7867

ISBN 0-408-01111-4

Typeset by Tunbridge Wells Typesetting Services Limited
Printed in England by Billing & Sons Ltd., Guildford, London and Worcester

Preface

The evolution of electronics technology has been founded on the fabrication of semiconductors and the consequent development and inauguration of the silicon integrated circuit.

Exploitation of the potential of the integrated circuit by the miniaturisation of electronics components has resulted in the proliferation of new products ranging from hand-held calculators to communications satellites and computers. This has necessitated the deposition of both passive and active components on a single silicon chip and has profoundly increased the capabilities of electronic devices.

To comprehend the interplay between component and circuit it is essential to understand the basic operation of such components. Hence the purpose of this book is to introduce the basic aspects of semiconductor components and to link the basic operating principles to the commercially available devices and systems.

The rate of progress in many technologies, especially those associated with microelectronics, is so rapid that textbooks tend to become obsolete almost as soon as they are published. It is hoped that by concentrating on fundamental principles and processes that are reasonably well established this book will avoid such a fate.

I have to express my thanks to the Editors of the following electronic journals who have very kindly granted me permission to reproduce the many articles I have contributed to their pages: *Electron, Electrical Review, Electrical Times,* and *Engineering Materials and Design.*

Acknowledgment is also made to Mullard Ltd, transactions and Proceedings of the Institute of Electrical and Electronics Engineers (New York) who through information contained in their technical reviews and journals have made possible a more complete presentation of the subject. The illustration on the cover of this book is reproduced by permission of the Digital Equipment Corporation.

Special thanks are also due to my wife for her valuable help in the preparation of the manuscript for publication.

Ilford   W. H. Dennis

Contents

1 BASIC ASPECTS 1
Electronics defined — Physical fundamentals — Atomic structures — Energy bands — Electrons and holes — Intrinsic and extrinsic semiconductors — p–n Junctions — Bias — Current-voltage characteristics — Passive and active components — Circuit elements — Printed circuits — Integrated circuits

2 MATERIALS 21
Metals — Semiconductors — Magnetic materials — Insulators — Ceramics — Plastics — Encapsulation materials

3 PASSIVE COMPONENTS — RESISTORS, CAPACITORS AND INDUCTORS 31
Resistance — Ohm's Law — Resistance and temperature — Resistor types — Thermistors and Variators — Capacitors — Inductance — Inductive reactance — Solenoids — Reluctance — Inductors — Transformers

4 ACTIVE COMPONENTS I: SEMICONDUCTOR DIODES 61
Rectification — Double-diode rectifier — Bridge rectifier — Zener diode — Gunn diode — Avalanche diode — Varactor diode — Schottky — Barrier diode — Tunnel diode — Unijunction transistor — Thyristor — Silicon controlled rectifier — Triac — Diac

5 ACTIVE COMPONENTS II: BIPOLAR TRANSISTORS 83
Introduction — *npn* type — *pnp* type — Transistor action — Current amplification — Carriers — Basic configurations — Common emitter — Common collector — Common base — Characteristics

6 ACTIVE COMPONENTS III: UNIPOLAR TRANSISTORS 99
Field-effect transistors — Basic types — Junction gate FET — Metal oxide semiconductor transistors — Enhancement type — Depletion type — Operation — Threshold voltage — Complementary MOS transistor — Characteristics — Charge-coupled device

7 FABRICATION OF SEMICONDUCTORS AND INTEGRATED CIRCUITS 118
Metal extraction — Crystal growth — Production — Alloy process — Planar process — Ion implantation — Gallium arsenide — Field-effect transistors Integrated circuits — Classification — Monolithic ICs — Future developments — Epitaxy — Hybrid circuits — Thin films — Thick films

8 ANALOGUE CIRCUITRY 143
General — Analogue circuits — Operational amplifiers — Voltage comparators — Voltage regulators — Analogue to digital converters — Filters — Impedance

9 DIGITAL LOGIC TECHNOLOGY 164
Binary system — Conversion — Logic circuits — Logic functions — NOT gate — AND gate — OR gate — NAND gate — NOR gate — MOS logic — Integrated-injection logic — Bistable circuits — Resistor-transistor logic — Diode-transistor logic — Transistor-transistor logic — Emitter-coupled logic

10 DIGITAL COMPUTERS 185
Hardware — Arithmetic and logic unit — Control unit — Memory — Solid-state memories — Magnetic bubble domain memory — Software — Machine language — High-level language — Algol — Fortran — Cobol — Coral — Input — Computer arithmetic

11 MICROPROCESSORS 204
General aspects — Circuit elements — Memory — Bipolar devices — Software — Assembly and high-level languages — Applications — Future trends — Microsensors

12 **VERY-LARGE-SCALE INTEGRATION** 216
Cost reduction — Size reduction — VLSI requirements — Available technologies — Fabrication — Alternative technologies — 64K-bit RAM — Role of silicon — Gallium arsenide — Josephson tunnel junction — Magnetic bubbles — Obstacles

13 **OPTOELECTRONIC COMPONENTS** 228
Photoelectronic devices—Photovoltaic effect — Solar cell — Photoemission — Photomultiplier — Photoconductivity — Photodiodes — Phototransistors — Photodetector arrays — Electroluminescence — Displays — Luminescence — Materials — Electrophoretic displays — Fibre-optics

Further reading 254

Index 255

Symbols, abbreviations and units

The symbolic notation used in the text conforms to conventional usage. Electrical units form part of the internationally adopted Systéme International d'Unités or SI units in which there are seven basic units, the metre (m), kilogram (kg), second (s), ampere (A), Kelvin (K), mole (mol) (amount of substance) and candela (cd) (luminous intensity).

Derived SI units

Quantity	Unit	Unit symbol
Electrical charge (Q)	Coulomb	C
Electrical resistance (R)	Ohm	Ω
Electrical current (I)	Ampere	A
Electrical potential (V)	Volt	V
Energy	Joule	J
Conductance (G)	Siemens	S
Impedance (Z)	Ohm	Ω
Inductance (L)	Henry	H
Capacitance (C)	Farad	F
Reactance (X)	Ohm	Ω
Power	Watt	W
Magnetic density (B)	Tesla	T
Magnetic flux ( )	Weber	Wb

Decimal multiple prefixes

In electronic systems numerical values of units may vary considerably in magnitude from very small to very large. For instance the cycle time required to execute an operation of a computer is measured in nanoseconds (1 n = 10^{-9} s) and capacitor values may be measured in millionths of a farad. For convenience in writing such values, prefixes are used to indicate decimal multiples and submultiples of the various units concerned.

Prefix	Symbol	Unit
tera	T	10^{12}
giga	G	10^{9}
mega	M	10^{6}
kilo	k	10^{3}
centi	c	10^{-2}
milli	m	10^{-3}
micro	μ	10^{-6}
nano	n	10^{-9}
pico	p	10^{-12}

Mega (M) and kilo* (k) are most usually associated with values of resistors and radio frequencies. Milli (m) and micro (μ) are most commonly associated with very low values of current and voltage. Capacitor values are invariably quoted in microfarads (μF) and picofarads (ϱF).

* Note that in computer work the symbol K is used since this does not represent exactly 1000 but rather 1024 or 2^{10}. Similarly, a 4K RAM would store 4096 bits of information.

Transistor symbols

Transistors are primarily concerned with the amplification of small signals which are conventionally represented by lower case (small) letters (v and i) while upper case (capital) letters (V and I) are used for amplitudes (peak values). Subscripts are attached for purposes of distinguishing transistor parameters.

V	d.c. or large signal voltage
V_I	Input voltage
V_O	Output voltage
V_b	Base voltage
V_c	Collector voltage
V_e	Emitter voltage
V_{CC}	Supply voltage
V_{ce}	Collector–emitter voltage
V_{cb}	Collector–base voltage
V_{eb}	Emitter–base voltage
v	a.c. voltage, small signal
v_e	a.c. voltage, emitter signal
v_I	Input voltage, small signal
v_O	Output voltage, small signal
I	Current
I_I	Input current
I_O	Output current
I_b	Base current
I_c	Collector current
I_e	Emitter current
I_{co}	Collector leakage
i	a.c. current, small signal
i_b	Base current, small signal
i_c	Collector current, small signal
i_e	Emitter current, small signal
n^+	High doping level
R	Resistance
R_c	Collector resistance
R_e	Emitter resistance
R_F	Feedback resistance
R_L	Load resistance
CB	Common-base circuit
CC	Common-collector circuit
CE	Common-emitter circuit
$\alpha(h_{FB})$	Current gain (CB)
$\beta(h_{FE})$	Current gain (CE)
FET	Field-effect transistor
I_d	Drain current
$I g_g$	Gate current

V_{ds}	Drain–source voltage
V_{gs}	Gate–source voltage
V_{th}	Threshold voltage
V_p	Pinch-off voltage
ω	frequency (rad s^{-1}) s
f	frequency (Hz)

1 Basic aspects

Electronics is essentially the science and technology of controlling the flow of electrons under the influence of applied electric or magnetic fields to produce useful results. The flow of electrons is controlled by means of electronic components which are concerned with the transmission, storage and processing of information whether it appears in electrical, optical or any other form. Components are usually thought of in terms of devices such as simple relays, switches or integrated circuits, but the ever-increasing complexity of systems means that computers can now be considered as components of even larger systems. Radio, television, radar, control of both production and quality in industry, and in fact whatever major field of social activity or world industry is selected it will be found that electronic devices play an increasing dominant role.

Physical fundamentals

Electrical current consists of the organised movement of charged particles which make up the atom, and the structure of the atom will be briefly considered here. A summary can best be presented by a discussion of the three major factors pertaining to an atom's basic electronic content. These are (1) atomic structure, (2) energy bands and levels and (3) electrons and holes.

Atomic structure

The atom consists of a positively charged nucleus around which negatively charged electrons circulate in a spatial orbit. Externally

2   Basic Aspects

all atoms appear electrically neutral in that the number of positively charged particles (protons) within the nucleus equals the number of negatively charged electrons. An element is positively charged when the number of electrons is less than the number of positive charges in the nucleus. When the number of electrons exceeds that of the positive charges, the atom is said to have a negative charge.

From this it is apparent that the electron is a most important particle. It not only represents the smallest carrier of electric charge but it also determines the type of charge of a given material as well as the magnitude of voltage and current.

A voltage is present whenever two points under comparison have differently charged atoms. For instance, if a point in a given circuit shows an excess of electrons in comparison to another point, a potential exists. The magnitude of this potential is a measure of the force which is necessary to separate electrons from their atoms. If two such points are connected by an electric conductor the electrons will flow from the negative point to the positive point until the potential difference has been equalised. This flow of electrons constitutes current.

Energy bands

The electrons surrounding the atom orbit at discrete distances in defined bands or shells. (Discrete literally means separately distinct.)

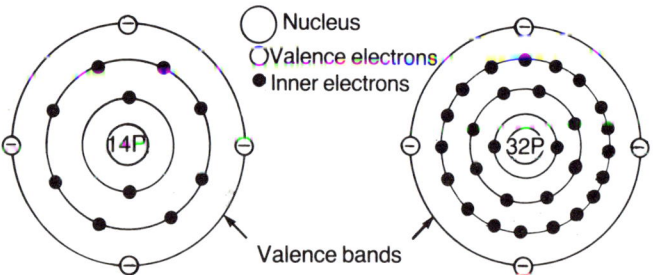

Figure 1.1. Silicon and germanium atoms. Silicon has a total of 14 electrons located in 3 bands or shells and 14 protons in the nucleus. Germanium has 32 electrons located in 4 bands and 32 protons in the nucleus. Each has 4 valence electrons which occupy the atoms' outer boundary and are responsible for the flow of current. The rest of the electrons form closed inner shells and are tightly bound to the nucleus

Because of their motion the electrons possess kinetic energy which determines the band followed by each electron. The greater an electron's energy the further it orbits away from the nucleus. The highest energy band is known as the valence band (Figure 1.1). A discrete level of energy in this band provides the force that binds all the electrons in the valence band of one atom to the electrons in the valence band of other atoms in the molecule. This is known as covalent bonding (Figure 1.2).

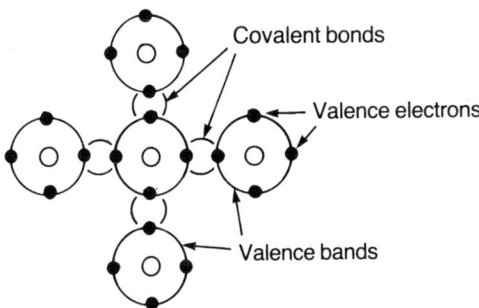

Figure 1.2. Two-dimensional representation of a silicon or germanium crystal the atoms being bound by covalent bonds. For simplicity only electrons in the valence bands are shown

The word valence, meaning chemical bond, indicates that the electrons are capable of linking up with neighbouring atoms and

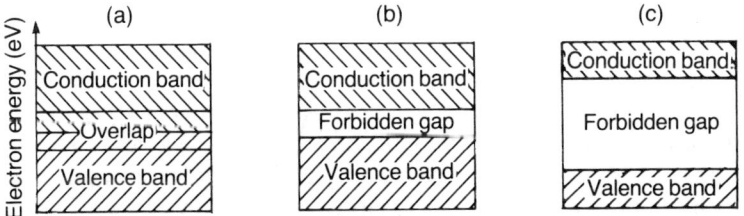

Figure 1.3. Energy level diagrams. Electrical conductivity depends on the spacing and state of occupancy of the electrons within the crystal. In a metal (a) the valence and conduction bands overlap and hence negligible energy is required to enable the electrons to move inside the crystal and conduct electricity. In semiconductors (b) the forbidden gap is small and it is not to difficult for some of the electrons to acquire the energy to jump across the gap. In an insulator (c) the gap is too large to be bridged and hence all the electrons are tightly bound to atoms and none is available to act as a carrier

thus of taking part in chemical or physical reactions. The next highest band is called the conduction band. This is only partially filled with electrons so that they can move freely in it, hence permitting conduction of current. The conduction band is separated from the valence band by an energy gap known as the forbidden band (Figure 1.3) within which no free electrons can exist. An input of energy can, however, cause a valence electron to jump across the gap into the conduction band. The forbidden gap varies in width in different materials. In the case of an insulator (c) the valence and conduction bands are so far apart that the gap is too large to be bridged by electrons. In a semiconductor (b) where there is a small energy gap between the valence and conduction bands, electrons can jump across the gap by a small expenditure of energy such as that represented by the voltage applied to a diode or transistor. In a conductor (a) the valence and conduction bands overlap and thus a transfer of electrons from one band to the other is easily established. Hence, it is apparent that the energy levels determine the physical nature of the material. The level of energy can be expressed in electron volts (eV), the general unit of moving particles. An insulator may have a level of several eV while the level for a semiconductor such as silicon is 1.1 eV and for germanium is 0.7 eV.

Semiconductors

Electrons and holes

When a negatively charged electron moves into the conduction band a corresponding imbalance is left in the valence band. This deficiency results in an electron vacancy and is known as a 'hole'. The hole left by the electron can be filled by another electron which in turn leaves behind another hole, and so on. These holes behave as if they were positively charged and in fact current is transported by their motion which can be regarded as electron motion in the opposite direction. In other words, electric conduction consists of negatively charged electrons and positively charged holes. Even though hole current is essentially due to the movement of electrons it is not the same type of electron motion as that which makes up the free electron current. The electrons which cause the hole current, jump from hole to hole and do not have enough energy to become substantially free.

This type of conduction by holes can only occur within the valence band for it depends upon the properties of the rest of the electrons in the band. It is unique to semiconductors and in fact hole conduction is fundamental to the operation of most types of semiconductor device.

Intrinsic semiconductors

In a highly pure semiconductor crystal of silicon or germanium (known as intrinsic) neighbouring atoms are bonded together by valence electron sharing. Covalent bonding firmly anchors the electrons rendering it difficult for an electron to break away from its bonds and participate in the transport of current. Hence such crystals would appear to be perfect insulators. However, this is true only at a temperature of absolute zero where there is no energy to agitate the electrons. At room temperature, owing to thermal effects, some electrons are set free (Figure 1.4) and move across the

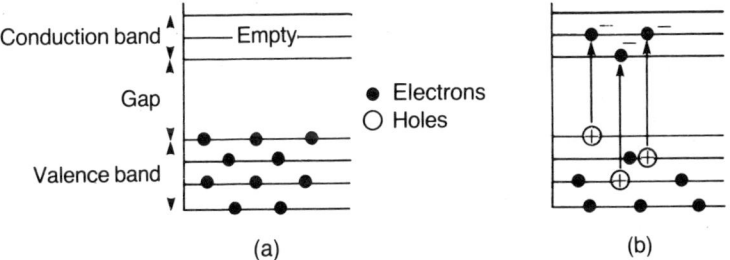

Figure 1.4. Energy-band structure of an intrinsic crystal. (a) At absolute zero the valence band is filled with electrons, and the conduction band is empty. (b) As temperature increases some electrons gain sufficient energy to break free from covalent bonding and move into the conduction band leaving behind an equal number of holes

energy gap leaving an equal number of corresponding holes in the valence band, giving rise to a small flow of conductivity.

Extrinsic semiconductors

The intentional addition of a very small amount of impurity, typically one impurity atom to every 10^6 atoms of an intrinsic semiconductor modifies the electrical characteristics to such an extent that the material becomes a much better conductor. Such material is called extrinsic since the condition results from the external addition of impurities, the operation being known as 'doping'.

Addition of impurity to an intrinsic semiconductor increases either the electron concentration in the conduction band or the hole concentration in the valence band.

As an example consider the doping of silicon which is a tetravalent element with four valence electrons. If an impurity element such as antimony or arsenic which is pentavalent is added; only four of the impurity's five valence electrons enter into covalent bonds with the silicon atoms in the crystal (Figure 1.5). The fifth electron does not enter into the covalent bonds but moves into the conduction band and acts as a current carrier.

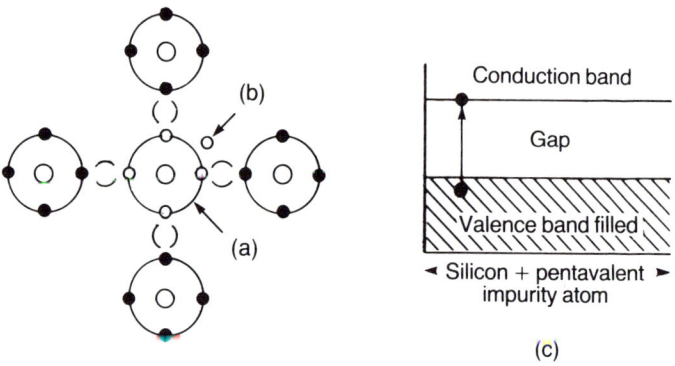

Figure 1.5. n-type semiconductor material (extrinsic). (a) Pentavalent impurity atom. (b) Free electron not bound by covalent bond. (c) The fifth valence electron of the impurity atom moves from the valence band into the conduction band and becomes available as a current carrier

A semiconductor crystal doped with a pentavalent impurity is called n-type since it contains free electrons which are negative charge carriers. Pentavalent impurity atoms are called *donor atoms* since each donates one free electron to the material. If, on the other hand, a trivalent element such as aluminium or phosphorus is used as the doping impurity there will be a deficiency of one valence electron (Figure 1.6). This creates a vacancy or hole which attracts an electron from one of the neighbouring atoms, and hence the presence of holes gives rise to conductivity. A semiconductor crystal doped with a trivalent impurity is called *p*-type since it contains positively charged holes (Figure 1.7). In this case the impurity atoms are called *acceptors* because each atom can accept one electron from the semiconductor material.

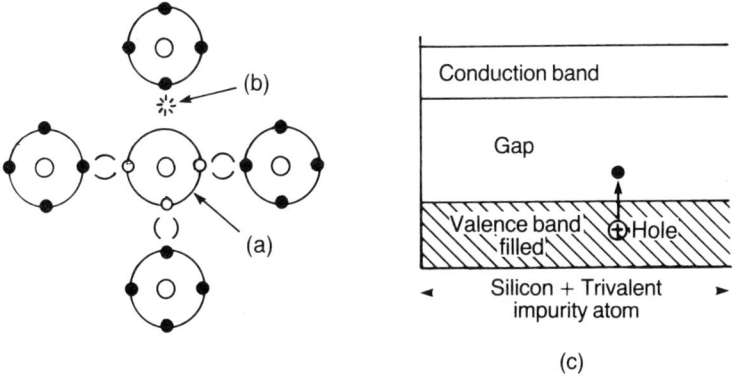

Figure 1.6. p-type semiconductor material. (a) Trivalent impurity atom; (b) hole; (c) addition of acceptor atoms increase the free-hole concentration in the valence band and hence the conductivity

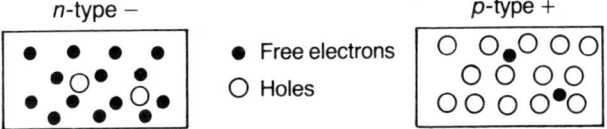

Figure 1.7. n- and p-type semiconductors. n-type has free electrons only at absolute zero but at room temperature a few thermally generated holes are present. p-type has holes only at absolute zero but at room temperature a few electrons exist in the free state owing to thermally ruptured covalent bonds

Junction of *p* and *n* semiconductors

If an *n*-type semiconductor is joined with a *p*-type material a *p–n* junction is produced (Figure 1.8). Note that this junction is a transition from *p* to *n*-type semiconductor material within a continuous crystal structure; merely to join physically *p*- and *n*-type material will not result in a structure having the electronic characteristics of a *p–n* junction.

Basic Aspects

There will be an excess of holes in the *p*-region and so these holes constitute what are known as majority carriers. A few thermally generated electrons will exist in this region and hence they are called minority carriers. Conversely in the *n*-region the electrons are the majority carriers and the holes the minority carriers. As soon as the junction is formed there will be a flow of carriers across it. Holes move into the *n*-type material and electrons move into the *p*-type material. The flow of electrons into the *p*-type make it negatively charged and similarly the flow of holes into the *n*-type make it positively charged. Since like charges repel, the negative charges oppose the further flow of electrons and the positive charges oppose the further flow of holes.

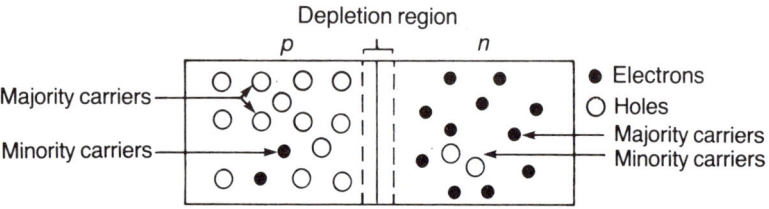

Figure 1.8. Diagrammatic representation of a p–n junction. When the p and n semiconductors are formed together to produce a junction, the majority carriers near the junction move towards each other and cancel out, a carrier-free area or depletion region being created

The majority charges therefore maintain positions away from the junction and consequently an effective potential gradient or barrier is produced prohibiting the passage of the current carriers. Hence in the area of the *p–n* junction the concentration of carriers (holes and electrons) on each side of the junction is much less than throughout the material and as a result this area is known as the depletion region because it is depleted of mobile carriers. This region is also known as the space charge region or barrier region.

Bias

Bias is a term frequently encountered in semiconductor technology and is fundamental in the operation of the *p–n* junction diode and the bipolar junction transistor. Literally it means to influence or dispose to one direction and as applied to semiconductors refers to the direction of current flow. When a voltage source is applied to the junction it is said to be biased and the applied voltage is called the bias voltage. Through the junction there is a forward or low-resistance direction and a reverse or high-resistance direction, the p–n junction acting as a one-way valve or rectifier to the flow of current. This is the characteristic of a semiconductor diode (chapter 5). Current flowing in the high-resistance direction is called reverse bias and current flowing in the low-resistance direction is called forward bias.

Reverse bias

If a voltage source such as a battery is applied to the junction as shown in Figure 1.9a the junction is said to be reverse biased. The

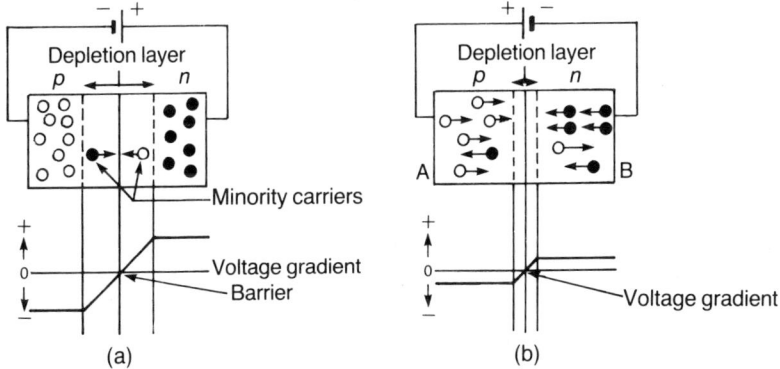

Figure 1.9. Biasing a p–n junction. (a) Reverse bias — depletion region is widened and the barrier is increased. The flow of majority carriers (holes and electrons) is prevented, only a very small current due to minority carriers takes place. (b) Forward bias — depletion region is narrowed and the barrier is reduced enabling majority carriers to cross the junction giving rise to large current flow

positive terminal is connected to the n-type material and the negative terminal to the p-type material. With the junction biased in this way the free electrons of the n-type are attracted to the positive terminal and the holes of the p-type are attracted to the negative terminal. Thus the majority carriers are pulled away from the junction, the depletion region being widened which has the effect of increasing the height of the barrier and as a result no majority carrier can cross the junction. Thus when a reverse bias is applied to a $p-n$ junction the only current that flows is the current due to the minority carriers.

Forward bias

When the positive terminal of the battery is connected to the p-type material and the negative terminal to the n-type (Figure 1.9b) the junction is said to be forward biased. In this case the holes in the p-type are attracted to the junction by the positive charge on terminal A and the free electrons in the n-type are attracted to the junction by the negative charge on terminal B which has the effect of decreasing the depletion layer. Electrons and holes move freely across the barrier, and hence a current flows across the junction consisting essentially of a flow of majority carriers, the action resulting in a current flow in the external circuit.

Summing up, reverse bias is a voltage applied to a junction in the high-resistance direction a voltage gradient being created. The greater the reverse bias the steeper the slope of the gradient and the larger the barrier. Forward bias cancels out the voltage gradient, overcoming the barrier.

Hence, it is apparent that the $p-n$ junction is the most important and essential part of semiconductor devices. A single $p-n$ junction forms the basis of semiconductor diodes, while the bipolar transistor uses two $p-n$ junctions. Silicon controlled rectifiers (thyristors) such as the triac have three junctions. These various types of components will be discussed in chapters 4 and 5.

The current-voltage characteristoc of a $p-n$ junction

A graph illustrating the relation between voltage and current characteristics in the behaviour of a silicon $p-n$ junction is shown in Figure 1.10. For comparison, that for germanium is also given. With forward bias, only a fraction of a volt need be applied to start a current. This initial voltage is of the order of 200 mV for germanium and 600 mV for silicon diodes. Once this voltage has been reached an extremely small change in voltage results in large changes in current. With reverse bias, because the $p-n$ junction allows current flow in one direction only, very little current flows. This small amount of current known as the reverse leakage current (I_0) is due to the fact that electrons and holes are continually being broken from some covalent bonds in the atoms by thermal energy to form minority carriers. This reverse leakage current is much less in silicon diodes than in germanium diodes.

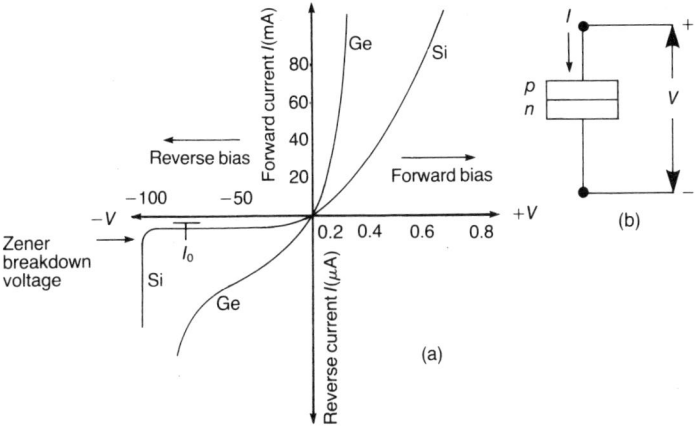

Figure 1.10. Current–voltage characteristic of a silicon p–n junction. (a) Current flow in the forward-bias direction is high and is measured in mA. Forward voltage is low. In the reverse-bias direction the leakage current (I_0) remains low (µA) until at a specific voltage — known as the Zener breakdown voltage — a sharp increase in current occurs. Germanium has a much higher leakage current when reverse biased. (b) The circuit diagram

Up to a certain voltage the reverse leakage current for silicon increases very little but after this voltage is reached a 'breakdown' occurs and a considerable flow of current results. This effect is caused by the minority carriers forming the reverse leakage current gaining sufficient velocity to dislodge other electrons from their covalent bonds thus increasing the number of free electrons and holes. There is, therefore, a sudden build-up of current carriers, the current suddenly rising from microamperes (µA) to milliamperes (mA) this effect being known as the avalanche effect. The voltage at which this sudden increase of current occurs is known as the Zener voltage or breakdown voltage. This does not imply a breakdown in the real sense of the word for no damage is done by operating in this region. This breakdown effect is exploited in voltage regulation devices known as Zener diodes (page 67).

Passive and active components

Electronic circuits include a variety of component parts known collectively as hardware. Hardware includes such components as resistors, capacitors and inductors (Figure 1.11) which are known as 'passive' components for they are not capable of supplying energy to the circuit of which they form part, but can only absorb or transfer signal power.

By contrast electronic semiconductor devices such as transistors and diodes which control, modulate or amplify the flow of energy in a circuit are known as 'active' components. These components are not themselves sources of energy but are able to supply it by abstraction from sources such as batteries or accumulators.

Passive components

Any material capable of carrying an electric current exhibits all the characteristics of passive circuit elements, that is resistance, capacitance and inductance. The various passive components, because of the differing nature of their composition, possess one of these attributes to a much greater degree than the others.

The electrical resistance of a metal can be interpreted as the friction to or hindering of electrons in motion. Another factor is the bond between the valence electrons and their nuclei. This bond is considerably stronger for resistance materials than for good conductors and hence the use of such high-resistance materials as carbon and nickel–chromium in the manufacture of resistors.

Capacitance and inductance are effects due to the electromagnetic field generated by an electric charge or current. The basic property of a capacitor is its ability to store electrical energy. In practice this is effected by an assembly of parallel metal plates separated by a thin insulating layer such as air, mica or plastic known collectively as dielectrics.

Inductance is represented by the build-up of a magnetic field by electrons moving in a conductor. To concentrate the magnetic field,

Passive and active components 13

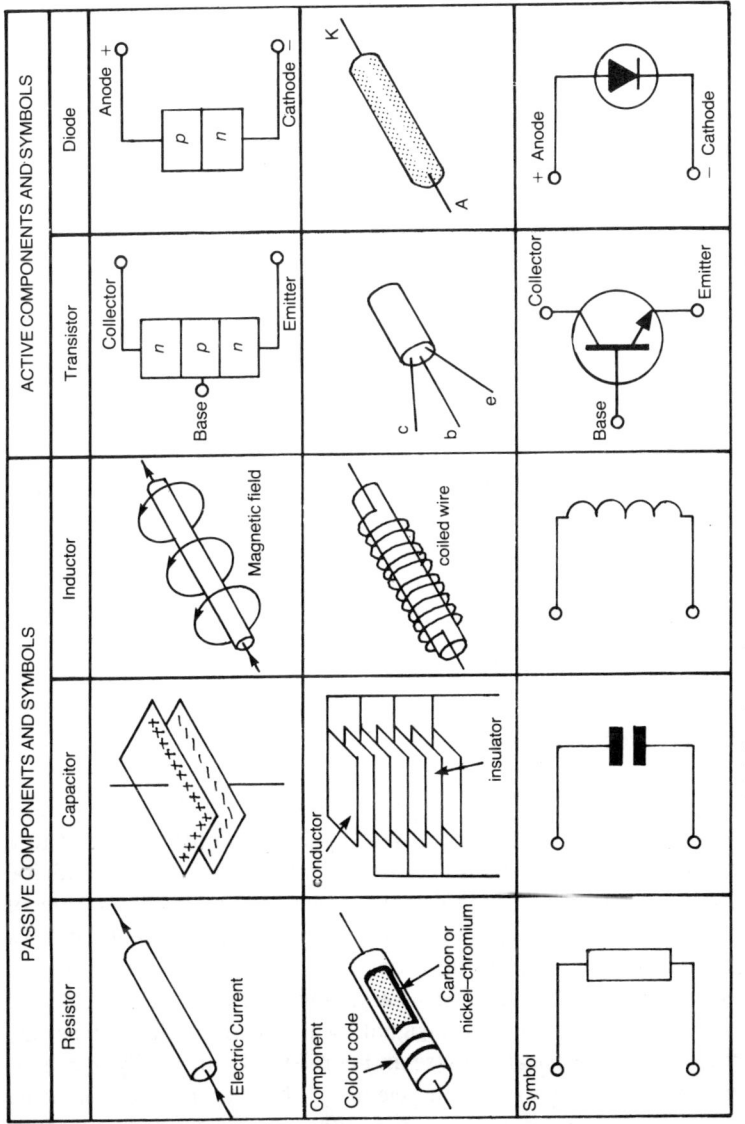

Figure 1.11. Passive and active components used in electronic circuits

inductors are made by winding the conductor in the form of a coil. Generally the coil is wound round an iron core which intensifies the lines of force resulting in a stronger magnetic field.

Active components

Passive devices are all symmetrical in the sense that their effects on a signal are the same no matter what the polarity of the signal. Active devices on the other hand are asymmetrical, the diode for instance depending for its action on its ability to pass current in one direction whilst limiting the flow of current in the opposite direction. In the reverse (blocking) direction the resistance may be very high, approximating to 5000–18 000 times that in the forward (passing) direction so that only a very small reverse current can flow. In this respect the transistor is also asymmetrical but it has the very important additional property that it is capable of amplification, an input signal of low power being converted into a higher power output. Thus a current gain or amplification of 200 and a voltage gain of 250 are common. This is the essential difference between active and passive components for the latter cannot increase the power of a signal and in fact the reverse takes place since the signal is always reduced.

Circuit elements

Much of the technological activity that concerns electronics relates to circuits which may range from a simple amplifying circuit to a complex one that performs mathematical calculations as in a digital computer. Typically an electronic circuit is any configuration of components and interconnections through which current can flow and whose function is the processing or generation of changing voltages and currents. When a varying voltage or current is used in a network it is known as a signal, which can be defined as any transmitted electrical impulse and consists of current in or voltage impressed upon a circuit element. In general, circuits are interconnections of both passive and active components such as resistors, capacitors and semiconductors which perform a required function. In circuit diagrmas the various devices are represented by symbols showing the major electrical property of the component. The more common are listed in Figure 1.11, and an example of an electronic circuit containing such devices is shown in Figure 1.12.

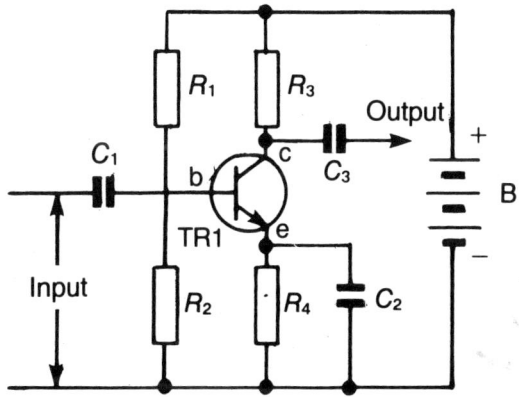

Figure 1.12. Amplifier circuit diagram illustrating the representation of components by symbols. R_1 and R_2, voltage divider; R_3, load resistor; R_4, emitter resistor. C_1 and C_3 blocking (d.c.); C_2, bypass. TR1 npn transistor amplifier. B, battery

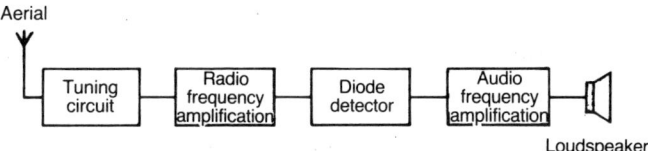

Figure 1.13. Simplified block diagram of a radio receiver

It is sometimes convenient to represent the signal processing that occurs in an electronic circuit by means of a block diagram. This consists of a diagram in which the important units of the signal processing operation are drawn in the form of interconnecting blocks usually rectangular in shape, the blocks being labelled according to the function they perform (Figures 1.13 and 1.14). No components are shown in the individual blocks, each block representing a separate sub-circuit.

16 Basic Aspects

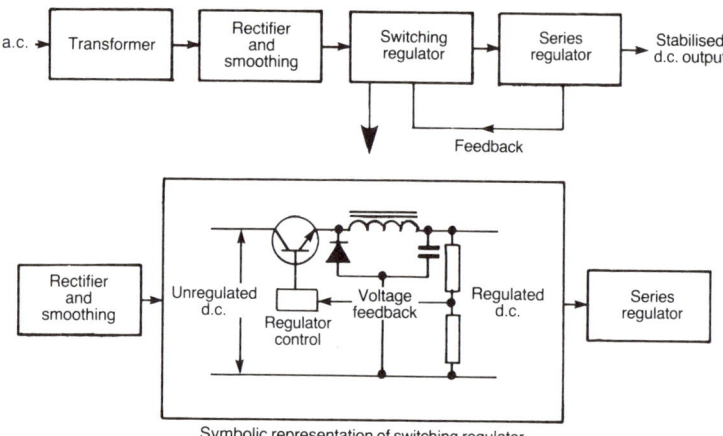

Figure 1.14. Block diagram illustrating conversion of alternating current to direct current. Output from the transformer is rectified and smoothed to give the required stabilised output

Printed circuits

A printed circuit comprises an interconnection by means of thin copper conductors laminated on to a plastics base board. In addition to providing the necessary interconnections, it also provides the mechanical support for the components. It can be easily mass produced by a photographic etching process, hence the term printed circuit.

Fundamentally a printed circuit functions as a substitute for wire leads connecting the component parts of electronic apparatus and arose from a need to evolve a cheap and economic method of physically wiring the components together to form the required circuit. It comprises a thin plastics board bonded on one or both sides with a thin layer of copper coated with a photosensitive material, which receives an imprint of the desired circuitry from a photographic negative of a printed circuit drawing. Acid resist is then applied and the board put into an etching bath. The etching solution eats away the copper except in the areas corresponding to the applied photographic imprint leaving the required circuitry as a series of thin strands of copper adhering to the plastics base. In this manner a replica of the printed circuit drawing is reproduced in copper. In addition to wiring, printed circuits can also embrace printed components such as resistors, inductors and capacitors. A logical extension of the rigid printed circuit board is the flexible printed wiring which has recently become available. The advantage is that while the stiffness of the rigid board is used to support mechanically the components mounted on it in one plane, the flexible circuit can be formed in three dimensions to interconnect points which do not fall on the same plane. They can be bent and folded along the inside of equipment to provide a three-dimensional printed wiring system. The range of applications is growing rapidly over the whole electronics industry and the technique is currently being used in colour television, telephone handsets, railways signal equipment, computers and flight instrumentation.

Improvements in the design and manufacture of printed circuit boards (PCBs) in recent years have made possible the miniaturisation of board components and the automation of design. Now, 50 times as many components can be put on an average board compared with a few years ago and the complexity has demanded automatic computer routing systems to keep the connection lengths to a minimum. The miniaturisation of components has led to the production of multi-layer boards, some with four or more layers. Another innovation is the production of thick-film hybrid circuits and to a lesser extent its companion the thin-film circuit. These are circuits that are deposited by screen printing on to ceramic baseboards and effectively form a bridge between the integrated circuit (IC) on the one hand and the PCB on the other.

Integrated circuits

An integrated circuit can be defined as a monolithic structure in which all the circuit components and connections are fabricated within or on a small chip of crystalline silicon. Formerly it was the practice to manufacture each of the components (transistors, diodes, resistors etc.) separately and then incorporate them into a circuit by wiring the components together with metallic conductors.

The transistor technology developed in the 1960s, particularly the silicon planar diffusion technology (chapter 7) made possible the fabrication of resistors and capacitors in a manner similar to the fabrication of the transistor. The development of passive components using the same fabrication technology and having similar space requirements as a transistor meant that all the circuit components could be made simultaneously on the same chip.

The substitution of ICs for discrete components offers the following advantages:

(1) A reduction in cost. Since thousands of components can be fabricated on one small slice of silicon (1.25 mm square) by inexpensive steps the cost is very small.
(2) A drastic reduction in size, volume and weight.
(3) Since ICs are much smaller they consume much less power than the components they have replaced.
(4) Since delay times are directly proportional to the dimensions of circuit elements, the circuit becomes faster as it becomes smaller.

It is no exaggeration to say that most of the technological achievements of the past decade have depended on the IC. The small size has been important in many applications ranging from communication satellites to hand-held calculators, digital watches and computers. The major impact, however, has been to make electronic functions more reproducible, more reliable and much less expensive.

Large scale integration
Large scale integration or LSI is a term applied to an extension of IC technology and refers to the miniaturisation of electronic components.

A LSI circuit may contain thousands of elements yet each element is so small that the complete circuit is typically less than 1 cm square. The scale may be gauged by enumerating the number of transistors contained in a chip.

Small scale (SSI) integrated circuits may have of the order of 10 transistors per chip; medium scale (MSI) circuits have of the order of 100. Microprocessor LSI circuits incorporate between 10 000 and 20 000 transistors, and in fact today's LSI chip carrying 10 000 transistors is comparable in size to a single transistor of 20 years ago.

Many factors have contributed to the growth in the number of circuit elements per chip. Improvements in the technique for growing large single crystals for the silicon wafers that contain the circuits have resulted in a doubling of the diameter of the wafer to 10 cm, hence many more chips can be made at one time. A further factor is the more efficient use of the silicon area leading to a hundredfold increase in the density of elements on the chip. These and other factors have led to the simultaneous fabrication of hundreds of circuits side by side on a single wafer and the manufacture of as many as 100 wafers together in a batch. For the most part LSI circuits are fabricated not so much with bipolar transistors but with metal-oxide semiconductor (MOS) transistors (chapter 6). Their importance in this area arises from many unique properties. First, because of the very small size it is possible to pack three to five MOSTs into the area occupied by one bipolar transistor. Secondly, it is possible to use them as high value resistors which also occupy a very small area. Thirdly, the very small capacitance that the gate electrode has to the rest of the structure can be used as the 'storage element' in memory circuits, the information being stored as charge in this capacitance. Hence the number of circuit functions that can be performed on a given chip of silicon is much higher using MOSTs. As a consequence circuit design using MOSTs is widely used for LSI, giving a significant decrease in costs compared with an IC using bipolar transistors.

Because large numbers of identical circuits are used in digital computers LSI is mostly found in computer and microprocessor systems. In this connection the major application for MOS technology has been in memory systems such as random-access, read only, and serial storage.

Very large scale integration (VLSI)
The concerted effort towards 'smaller and smaller, more and more on a chip' has ushered in VLSI as the dominant microelectronic technology of the next decade. VLSI circuits with densities of 50 000 components per square centimetre are already in high-volume production. The drive for greater component density is stimulating research into submicron techniques to avoid the optical diffraction limits imposed by the contact photolithography method of circuit patterning (chapter 7). These areas of development include direct writing by electron beams, X-ray lithography and automated wafer production. As a result, circuit densities of up to half a million

transistors on a single chip are foreseen. It would seem that VLSI is forming the foundation in pushing forward the frontiers of the amazingly versatile field of solid-state integrated circuits. The latest developments in VLSI are discussed in chapter 12.

2 Materials

Materials used in the manufacture of electronic devices range from metals, semiconductors, insulators and magnetic materials to ceramics and plastics. Metals such as copper and aluminium are used for establishing electrical contacts for electrical interconnections. Metallic compounds such as indium, antimony and arsenic are used in semiconductors as the source of intentionally added impurity atoms. Examples of semiconductors include such materials as silicon and germanium which form the basis of the modern electronics industry. Insulators are used for energy storage, charge storage, electrical insulation, for passivation of surfaces and as host materials for lasers. Two further classes of materials are the magnetic materials and the ceramics. The magnetics are used not only as straightforward magnetic components in computer memory units but also in components which combine electrical and magnetic phenomena for electronic functions. Ceramics are essentially inorganic non-metallic materials and in general are hard, brittle, resistant to heat and electrically insulating. They have special electrical and magnetic properties which render them useful as magnets, dielectrics and piezoelectric materials.

Metals

The use of metals in the electronic industry covers a wide range and only a few features can be selected for discussion. Apart from their desirable electrical and magnetic properties, they have in general excellent mechanical strength and ductility which may be modified by suitable physical treatment. A variety of metals are utilised in forming the passive components of electronic circuits. The principle functions performed by metals are: (1) transmission of electrical current from one compound to another; (2) as the external leads, mechanical supports and heatsinks for the dissipation of heat from transistors and diodes; (3) resistor, capacitor and inductor material. Copper has traditionally been used as conductivity and contact material becuase of its low resistance and ease of fabrication. Aluminium, because of its good electrical and mechanical properties, availability and low cost has in recent years achieved wide usage. Other more expensive metals are selected for stability and inertness with less concern for cost because so little is used in a single device. Gold and silver have excellent conductivity and may be advantageous in some applications due to superior corrosion and oxidation resistance at ambient and elevated temperatures. Purity is extremely important where the metal contacts the semiconductor component since even a low percentage of undesirable elements causes the formation of deleterious compounds and deterioration of device characteristics.

Semiconductors

In the thermionic valve, whose invention in 1906 heralded in the age of electronics, electric current flows as a stream of electrons between two or more electrodes contained in an evacuated glass bulb (Figure 2.1). The name thermionic derives from the fact that heating (electrically) is the source of the electrons. They are costly, bulky, complicated to manufacture and wasteful of power and because of these disadvantages have been supplanted by the transistor which was invented in 1948. In this device current still flows between electrodes but in this case through a solid material made from materials known as semiconductors.

Semiconductors when they are pure and at low temperature can be classed as non-metals for they do not conduct electricity. To obtain the controlled flow of current required in transistors, diodes and other electronic devices, semiconductors are made to conduct by introducing controlled amounts of impurity atoms into the crystal.

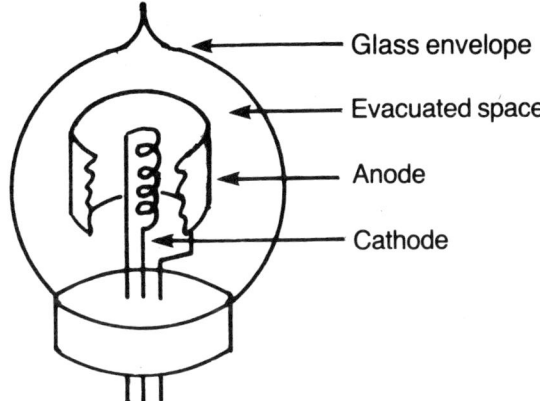

Figure 2.1. Thermionic valves were the basis of electronics before the transistor was invented in 1948. Nearly the size of a small electric light bulb, many thousands of such valves were used in the construction of a computer. Kiloamperes of heater current had to be provided and the whole device filled a large room

The impurity simply adds readily available electrons to serve as current carriers and since only a comparatively small numbers of carriers is required in a conductor, a minute amount of impurity can raise a pure semiconductor's conductivity to nearly that of a metal. Hence it is the creation of a built-in field by the introduction of impurities that makes possible the production of electronic devices from semiconductors.

In the early development of semiconductor materials in the 1950s germanium was the most important. However, by 1960 it was clear that silicon was destined to replace germanium in almost all aplications. The most significant advantages of silicon are that it has a lower leakage current, operates at higher temperature and is less affected by temperature changes because of the favourable properties of its oxide. This last property is most important in the fabrication of transistors etc., for it enables a dense, uniform insulating oxide to be grown on its surface which is stable over a wide range of temperature. This makes it possible to use the oxide as a mask, which while allowing the required dopant to penetrate the silicon through 'windows' prevents their admission elsewhere. By an extension of the technique known as the planar process (p. 123) several hundred transistors, diodes and other components can be simultaneously diffused into a single silicon chip about 5 mm × 5 mm (Figure 2.2) thus allowing the fabrication of integrated circuits.

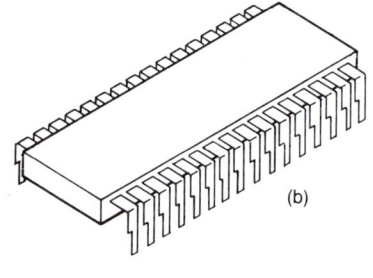

(a)

(b)

Figure 2.2. As technology developed, valves were replaced by transistors. Minicomputers, for instance, having the equivalent of 10 000 transistors are fabricated on this 5 × 5 mm silicon chip (a). (b) The encapsulated chip ready to be plugged into a circuit board

In addition to silicon and germanium many other elements are utilised in the manufacture of semiconductor materials. These are listed in Table 2.1 and are conveniently divided according to the

Table 2.1 Semiconductor elements

Group II 2 valence electrons	Group III 3 valence electrons	Group IV 4 valence electrons	Group V 5 valence electrons	Group VI 6 valence electrons
Beryllium	Boron	Carbon	Phosphorus	Sulphur
Magnesium	Aluminium	Silicon	Arsenic	Selenium
Zinc	Gallium	Germanium	Antimony	Tellurium
Cadmium	Indium	Lead	Bismuth	

number of valence electrons they possess. Group IV contains the two most important semiconductors, namely silicon and germanium, Group V contains the elements commonly used for doping germanium and silicon to give n-type semiconductors, whereas the elements in Group III yield p-type semiconductors. In addition to the elemental semiconductors there are other types of semiconductor materials in common use. These are known as compound semiconductors and are obtained from a combination of Group III and Group V elements or Group II and Group VI elements.

Examples are gallium arsenide, indium antimonide, cadmium and zinc sulphide. These tend to be found in specialist devices such as those used in the emission or absorption of light. For example, light-emitting diodes (LEDs) are made of gallium arsenide or gallium phosphide and electroluminescent material such as zinc sulphide is used on television screens.

Semiconductor characteristics

The properties upon which semiconductors depend for their action are as follows:

(1) In a pure (intrinsic) semiconductor, electrical conductivity is temperature dependent rising exponentially with temperature.

(2) At normal temperature the addition of a trace of impurity is required to ensure conductivity.

(3) In an impure (extrinsic) semiconductor the conductivity depends on the impurity concentration. The amount required is of the order of 1 part in 10^7 which increases the conductivity by a factor of 10^{10}.

(4) Depending on the type of impurity added, current flow may be either by electrons or holes.

(5) Both types of charge carrier are commonly present, transistors and diodes being examples of this type of semiconductor.

The important properties of semiconductors thus depend on the addition to the pure crystal of minute amounts of impurities. It is

Table 2.1

Property	Si	Ge	GaAs
Atomic numbers	14	32	33
Atomic weight	28.08	72.6	144.6
Specific gravity	2.33	5.47	5.32
Melting point (°C)	1370	958	1240
Energy gap (eV)	1.11	0.67	1.4
Thermal conductivity (W cm^{-1} K^{-1})	1.41	0.61	0.45

apparent that for this to occur the semiconductor crystal must in the first instance be in an extremely pure form. In fact semiconductors are probably the purest material available, impurities of 1 part in 10^{12} of material being commonly achieved. The method by which such purity is made possible is described in chapter 7.

Magnetic materials

Magnetic materials in general are distinguished by their characteristic magnetism which is the property of being susceptible to the action of a magnetic field. This property shows itself in different forms and with different intensities and materials can be classified on this basis into three groups:

(1) Diamagnetic. Susceptibility is negative, that is the magnetisation opposes the magnetising force, the permeability being less than unity.
(2) Paramagnetic. Susceptibility is small, the relative permeability is slightly greater than unity.
(3) Ferromagnetic. Has a high value of susceptibility, the relative permeability being much in excess of unity.

Alloys based on one or more of these ferromagnetic materials make up most of the range of magnetic materials used in the electronics industry, iron, cobalt and nickel alloys being the main ones.

From the technological viewpoint magnetic materials are either 'soft' meaning easy to magnetise or demagnetise, or 'hard' meaning the opposite. Soft magnetics are characterised by a high value of permeability and low coercivity. Permeability is a measure of the conductivity of magnetic flux through a material and is the ratio of magnetic flux density to the magnetic flux producing it. Coercivity is the reverse magnetising force required to 'coerce' a material back to zero induction after it has been saturated. It may range in value from a few ampere turns per metre in soft magnetics to thousands of ampere turns in hard magnetics. Soft magnetics are used for transformer cores found in television receivers, magnetic amplifiers and automatic controls.

Hard materials come into play where permanent magnets are required for use in loudspeakers, moving coil instruments and so on. A typical example is the alloy Alnico which is an alloy of aluminium, nickel, cobalt and copper, the name being derived from the chemical symbols of the first three metals.

Ferrites

Ferrites are ferromagnetic materials based on iron oxide and possessing exceptionally high coercivity which means that greater use can be made of available energy. Their electrical resistivity is also very high which makes them immune to eddy currents.

Some applications call for properties that lie halfway between hard and soft. Magnetic memory elements in a digital computer must be hard enough to retain their forward or reverse magnetisation — the states corresponding to 0 and 1 in the binary system — indefinitely on being stored. They must also be soft enough to switch states rapidly when a small external field is applied in the course of information read-in or read-out.

Insulators

These materials depend for their properties on the fact that the bonding of the electrons is so firm that they are not free to move in an electric field, unlike the electrons in metals which are mobile and act as the agents of conduction. Insulators, often referred to as dielectrics, are used for energy and charge storage as in capacitors, to isolate electrical elements and devices and also serve as substrates on which electronically active materials may be deposited. These substrates include glass, quartz, silicon monoxide, silicon nitride, sapphire and spinel.

Development of optical modulators and solid-state lasers has increased the demand and has stimulated methods for the growth of bulk crystals of insulators. Aluminium oxide (ruby and sapphire), titanium oxide, strontium titanate and calcium tungstate have all been grown by the Czochralski technique using extremes of temperature, pressure and controlled atmospheric conditions. The insulators are getting more attention as electronic interest swings

Ceramics

Essentially a ceramic may be defined as a combination of one or more metals with a non-metallic element usually oxygen. In general they are hard and brittle, electrically insulating, refractory, stable and inert. They have long served as electrical insulators, porcelain being perhaps the best known, formed by firing a mixture of china clay (Al_2O_3, SiO_2, $2H_2O$) and potash feldspar (K_2O, Al_2O_3, $6SiO_2$) at about 1300 °C. Ceramics now play more active roles based on special electrical and magnetic properties. The most important of their electronic applications are summarised in Table 2.3.

Table 2.3

Applications	Ceramic materials
Insulation	Steatite, alumina
Capacitors	Steatite, rutile, titanates
Semiconductors	Silicon, germanium, silicon carbide
Computer memory units	Barium titanate, lead zirconate titanate
Magnetic components	
Permanent magnets	Barium ferrite
Induction and transformer cores	Manganese–zinc and nickel–zinc ferrites
Piezoelectric devices	Barium titanate, lead zirconate titanate

An unusual property of certain ceramic materials notably quartz and barium titanate is that of piezoelectricity (pressure electricity). This involves the expansion along one axis of a crystal and contraction along another when subjected to an electric field and the converse effect whereby mechanical strain produces opposite charges on different faces of the crystal. Devices developed to convert mechanical energy into electric energy and vice versa are known as transducers and are usually made in the form of a plate or disc composed of quartz or lead zirconate titanate. The piezoelectric property makes it possible to excite and detect the mechanical oscillations of the crystal electrically, hence they are used for precise control of the frequencies of oscillators and for the generation of high-frequency sound waves. A common use of the property is in record player pick-ups which consist of two thin strips of piezoceramic glued together. The stylus or needle of the pickup follows the grooves in the record, the mechanical vibrations of the needle being transformed by the piezoelectric device into electrical voltages which are amplified by a voltage amplifier before being delivered to the loudspeaker. The converse piezoelectric effect is used in underwater sound generation (sonar).

Plastics

Plastics include those products manufactured from polymeric materials and denoted by their ability to be moulded by the application of heat into solid articles of any desired shape. Indeed the greatest single advantage of plastics is that they can be moulded into the finished article at a relatively low cost compared with the machining and fabricating necessary with wood and metals. hence the popularity of moulded plastics such as polypropylene which gives an excellent surface finish for the casings of radios, television sets etc. In general they are resistant to chemical corrosion, do not rust like iron, possess high strength–weight ratio, and their excellent dielectric properties single them out for use in capacitors. The polyester plastics, polyethylene, terephthalate and polycarbonate are well established as dielectric materials in capacitors. These dielectric materials are characterised by an insulation resistance which is higher than that of any other dielectric and falls relatively slowly with temperatures up to about 70 °C.

Encapsulation materials

Electronic components such as resistors, capacitors and wound components (inductors and transformers) depend for their performance on the maintenance of electrical potentials across dielectric media. Although current conductors are normally of metal and relatively stable, exposure to hostile environments adversely affects their operation. Application of a protective coating is therefore necessary and although this may introduce useful secondary characteristics such as strength and shock resistance, exclusion of moisture is of paramount importance.

The traditional materials such as waxes, bitumen and drying oils have now been largely replaced by plastics. A wide variety of materials is used for this purpose including polyurethanes, polyesters, PVC and silicones but the most widely used are the epoxies.

The desirable electrical properties, high volume resistivity and low dissipation factor as well as high mechanical strength which is retained under wet and humid conditions are features which reinforce the success of epoxies in electronic devices. In addition to low moisture absorption and permeability the epoxies possess excellent bonding characteristics and this adhesive property commends them for precision encapsulation.

Current practice

The actual method of encapsulation for protection against mechanical and environmental conditions is carried out by a number of techniques depending on the application requirements of the equipment in which they are used. Thermosetting mixtures, that is those plastics which become permanently hard when heated above a certain temperature, may be used cold in liquid form and then set by heating. Alternatively they may be compounded into moulding powders and applied under heat and pressure. Thermoplastics on the other hand which soften or liquify when heated and harden upon cooling are also in common use.

For improved reliability recourse to hermetic sealing is necessary. The equipment used in this technique comprises an upper heated chamber which contains the liquid plastics and a lower chamber in which the component or assembly is suitably mounted in a mould. Both chambers are capable of evacuation. When the desired vacuum is attained a valve is opened which allows the encapsulant to flow from the upper chamber into the moulds. The vacuum is then broken and the moulds removed to an oven to cure the plastics. The complete casting cycle takes about 3 minutes and as many components can be encapsulated as can be accommodated in the lower chamber.

3 Passive components — resistors, capacitors and inductors

Electronic circuits can be said to possess three properties, resistance (R), inductance (L), and capacitance (C). Resistance is an inherent opposition to the flow of electrons present in conductors. Inductance is the property of any circuit to oppose any change in current, whereas capacitance is the property of a circuit to oppose any change in voltage.

Resistance determines the current produced by a given potential difference and acts to limit current and to produce voltage drop to some desired level. Two or more resistors in series divide the voltage whereas when connected in parallel they divide the current.

Capacitors provide an impedance to electric flow inversely proportional to the frequency of the applied voltage. Principal uses are for tuning resonant circuits, for blocking d.c. voltages from an electrical circuit while permitting the passage of a.c. voltages, for by-passing or short-circuiting alternating voltages and as filters.

Inductors limit alternating current but permit the passage of direct current and find important applications in converting sinusoidal voltage to another of different amplitude. They are used to process electronic signals and in electrical engineering for handling power.

Resistance

Opposition to current flow is known as resistance, and can be defined by Ohm's Law which states that the current in a circuit is directly proportional to electromotive force and inversely proportional to resistance. It is measured in ohms the symbol for which is Ω. A resistance of 1000 Ω is denoted as 1 kΩ and one million ohms as 1 MΩ. Hence 1 000 000 Ω = 1000 kΩ = 1 MΩ. Materials show a great variability in resistivity and in fact it is one of the most widely varying of all physical quantities.

Table 3.1

Material	Resistivity (μΩ cm)
Aluminium	2.45
Copper	1.56
Gold	2.4
Iron	8.9
Mercury	94.1
Nickel	10.5
Silver	1.51
Graphite	600–1200

Metals such as silver and copper have very low resistance, iron, steel and nickel have a measurable resistance, whereas graphite, ceramics, wood and glass offer such a high opposition to current flow that their resistance can be considered to be virtually infinite and in fact they can be regarded as insulators.

Ohm's law

Resistance is related to current and voltage for the amount of current flowing in a circuit is dependent on the amount of applied electrical pressure (voltage) and the resistance encountered. The relationship between the three is given by Ohm's Law

$$I = \frac{V}{R}$$

which indicates that the amount of current flowing in a circuit is equal to the voltage divided by the resistance. Thus if the current from a 10-V battery flows through a 5-Ω resistor the current flowing through the resistor is 2 A.

By rearrangement of the formula, unknown values of either resistance or voltage can be ascertained. Resistance, for example, can be obtained if the value of current and voltage are known. Then

$$R = \frac{V}{I.}$$

Thus if the voltage is 10 V and the current is 2 A, a resistance of 5 Ω is indicated.

Hence it is evident that in electronic circuits, the amount of current depends not only on the voltage but also on the resistance.

Resistivity

The specific property that determines the resistance that a material offers to the flow of current is known as resistivity. It is denoted by ϱ (rho) and depends on the material from which the resistor is made. If R (Ω) is the resistance, l (cm) is the length of the resistor, A (cm²) the cross section then

$$R = \frac{l}{A} \varrho$$

when the resistivity (ϱ) is in μΩ cm at 0 °C.

Resistance and temperature

The electrical energy absorbed by a resistor must by the law of conservation of energy be converted into some other form. The most common is heat energy. With metals such as copper, iron and nickel, resistance increases with increase in temperature. On the other hand, the resistance of carbon and insulating materials such as rubber and paper decreases with temperature. In certain alloys such as Manganin which is a copper-base alloy (containing manganese and nickel) and Nichrome, a nickel–chromium alloy, the resistance remains practically constant. Although certain materials such as thermistors and varistors (p. 40) are made specifically to exploit the effects of temperature on resistance, for normal applications a significant change in resistance with temperature is not desirable.

Temperature coefficient of resistance (t.c.r)

This parameter signifies how the resistance changes with

temperature. At moderate temperatures the resistance R_t at $t\,°C$ is related to the resistance R_0 at $0\,°C$ by the formula

$$R_t = R_0(1 + \alpha t)$$

where α is a constant of the material known as its temperature coefficient. This coefficient is the fractional increase in resistance from $0\,°C$ when its temperature is raised by $1\,°C$. Its magnitude for most metals is about $0.004\,°C^{-1}$. The following will illustrate the use of the formula. If a copper coil has a resistance of $100\,\Omega$ when cold what will be its resistance if the temperature is raised by $40\,°C$?

$$R_{40} = 100(1 + 40 \times 0.004) = 116\,\Omega$$

Temperature coefficient of resistance in practical terms is generally given in parts per million and falls within the range 15–200 p.p.m. $°C^{-1}$.

Power rating

Flow of current through a resistor causes a rise in temperature and since all materials used in their construction have a maximum temperature above which they deteriorate, it follows that all units have a limited wattage dissipation. Thus resistors normally have a power rating which should not be exceeded. Power rating is given by $P = V^2/R$ where P is in watts, V in volts and R in ohms. As the majority of electronic circuits work on low voltages, resistors with moderate power ratings are usually adequate.

Superconductivity

In connection with temperature and resistance one of the most interesting of transformations is the total disappearance of resistivity in certain alloys (niobium–titanium and niobium–zirconium) when they are cooled to a temperature near absolute zero ($-273\,°C$). Once energised, current continues to flow without further input of energy. The complex and bulky refrigeration equipment required to maintain the alloys in the superconductive state is the main drawback to its industrial usage.

Types of resistor

To match the various requirements of their many applications resistors vary widely in size and composition. Although in general resistors are cylindrical in shape (Figure 3.1) with a lead at either end, their physical size ranges from types in which the body of the resistor is scarcely larger than a pinhead to types with overall dimensions of several centimetres. Materials used in their construction also vary and they may be considered under three main categories: composition, film and wirewond reactors.

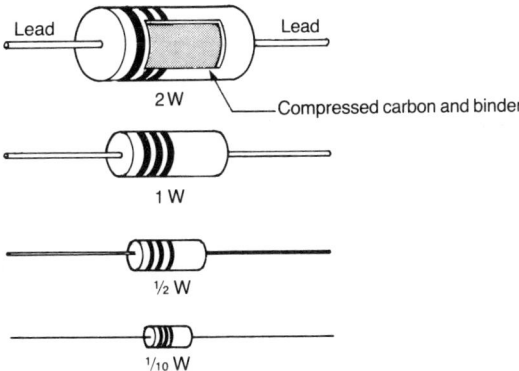

Figure 3.1. Composition (carbon) resistors (drawn to size). Such resistors cover the range 10Ω to 10 MΩ

Composition type

These are generally made with carbon as the conducting material and are used in great quantities, are very low in cost and are manufactured by two different moethods. (a) Carbon powder is mixed with a resin binder and some refractory material and then baked in a kiln. They may be made in the form of a ceramic tube or in the form of a moulded carbon rod. Wire leads are then attached to facilitate insertion into the circuit and they are then painted with a colour code. Varying the proportion of carbon can produce resistors ranging from a few ohms up to millions of ohms. They are usually of low wattage, starting at $\frac{1}{10}$ W and continuing through intermediate sizes up to 2 W (Figure 3.1). Such resistors are produced by the million for use in radio and communications and although they maintain a reasonably constant resistance value over a considerable temperature range they are not usually made to a very high order of precision. (b) A higher degree of stability is provided by the deposited carbon method. In this type a thin film of carbon is deposited on a ceramic or glass core either by applying a carbon coating and then heating or by heating in the presence of a carbon-containing gas such as methane whereby decomposition takes place.

Metal film resistors
These usually consist of metal or metal oxide deposited on a ceramic substrate. They are relatively inexpensive and because of their superior power dissipation and performance are extensively used in commercial applications. They are available with a t.c.r. in the range 300–500 p.p.m. °C^{-1} and a stability figure of 1–2% for 100 hours at full power rating. Although giving a superior performance compared with other film elements, power ratings are confined to 3 W and this tends to limit their application. They are made by a similar technique to carbon deposition, namely by decomposition of the appropriate metal chloride on a ceramic substrate by application of heat.

Wirewound resistors
These consist of wire alloy wound between terminals on a supporting ceramic former and suitably protected against mechanical damage (Figure 3.2). Typically the wire alloys used for resistors are

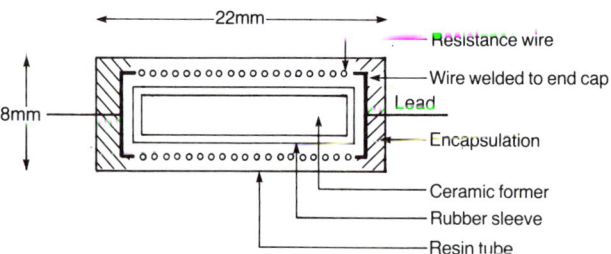

Figure 3.2. Wirewound resistor. Resistance tolerance ±5%; wattage rating 20°C) 16 W; resistance range 4–140 kΩ; temperature coefficient 40–100 ppm °C^{-1}

Nichrome (80–90% nickel and 20–10% chromium) and Manganin (86% Cu, 20% Ni, 12% Mn). When exceedingly high values of resistance (above 10 MΩ) are required a thick film of semiconductive glass such as borosilicate is fused to a ceramic substrate.

Colour code

It is common practice to provide coloured markings which denote the resistance value of each unit. In one code (Figure 3.3) the colour

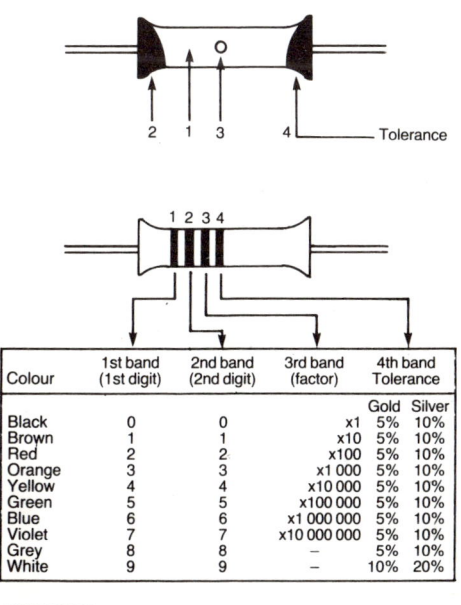

Colour	1st band (1st digit)	2nd band (2nd digit)	3rd band (factor)	4th band Tolerance	
				Gold	Silver
Black	0	0	x1	5%	10%
Brown	1	1	x10	5%	10%
Red	2	2	x100	5%	10%
Orange	3	3	x1 000	5%	10%
Yellow	4	4	x10 000	5%	10%
Green	5	5	x100 000	5%	10%
Blue	6	6	x1 000 000	5%	10%
Violet	7	7	x10 000 000	5%	10%
Grey	8	8	–	5%	10%
White	9	9	–	10%	20%

EXAMPLES
10 Ω   brown black black
100 Ω   brown black brown
470 Ω   yellow violet brown
1000 Ω   brown black red
2200 Ω   red red red
3300 Ω   orange orange red
10 000 Ω   brown black orange
100 000 Ω brown black yellow
470 000 Ω yellow violet yellow

Figure 3.3. Colour code for resistors (and capacitors)

of the body of the resistor designates the first digit. The colour of one end of the resistor designates the second digit and the colour of the dot or band in the centre indicates the number of ciphers after the first two digits. Tolerance is denoted by colour at the other end. In the other common colour code a series of bands at one end give the first and second digits, the factor and the tolerance (Figure 3.3). The colours used to designate different numbers are as follows:

Black	0	Blue	6
Brown	1	Violet	7
Red	2	Grey	8
Orange	3	White	9
Yellow	4	Gold	± 5% tolerance
Green	5	Silver	± 10% tolerance

Thus a resistor banded with green, black, red and silver is 50×10^2 Ω that is 5.0 kΩ with a 10% tolerance. In a more recent code (BS 1852) the value and tolerance of resistors are indicated by using letters and figures instead of colours. The multiplier is represented by a single letter, its position indicating the decimal point.

$$R = \times 1$$
$$K = \times 1000$$
$$M = \times 1\,000\,000$$

A second letter indicates the tolerance thus

$F = \pm 1\%$, $G = \pm 2\%$, $J = \pm 5\%$, $K = \pm 10\%$, $M = \pm 20\%$

Examples of BS 1852 resistor descriptions are

Older Code	BS 1852
0.53 Ω ± 20%	R53M
4 Ω ± 10%	4R0K
2.9 Ω ± 5%	2R9J
4 KΩ ± 10%	4K0F
49 KΩ ± 10%	49KK
6.6 MΩ ± 20%	6M6M

Tolerance

Although a resistor is designed to have a specified value of resistance an exact value is difficult to attain. To comply with this range in variation a percentage tolerance is usually introduced which indicates the permitted range above or below the normal resistance value. These are of the order of 1, 2, 5, 10 or 20%. Thus the actual resistance of a resistor of 100 Ω with a 10% tolerance would be anything between 90 and 110 Ω. Resistance values tend to change with time and hence stability is another important factor. A typical value of resistance stability (ΔR) would be 2% after 2000 h.

Variable resistors

It is often necessary to vary the resistance of a resistor which is permanently connected in a circuit. Such variable resistances use a mechanical slider which rides over the resistance element thus selecting the length of the element to be included in the circuit. In a tap type resistor the tap contact is positioned as desired on the resistance element and then tightened by a screwdriver. Connection can be made to the movable arm and to one end of the resistance element. In this event the unit is known as a rheostat. If connection is made to the movable arm and to both ends of the resistance element the unit is known as a potentiometer (pot). Hence a potentiometer is a three terminal resistor with an adjustable sliding contact that functions as an adjustable voltage divider and hence makes it possible to mechanically change the resistance. The size and rating is specified by giving its total resistance in ohms and the permissable losses in watts. It is used to adjust and control the electric potential difference (voltage) applied to some device or part of a circuit. It is also used as volume control in radio, instrumentation, process

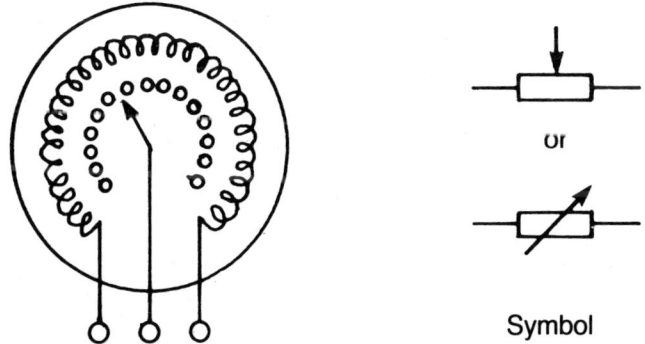

Figure 3.4. Wire-wound variable resistor (potentiometer). Resistance range 100–1 MΩ; resistance tolerance ±10%; rated dissipation 0.5 W at 70 °C

control panels and servo systems. In addition to the rectilinear type, rotary pots (Figure 3.4) are also available.

Thermistors and varistors

In a normal resistor the current remains proportional to the applied voltage. Components for which this proportionality does not hold are known as non-linear, and special devices have been developed which take advantage of this non-linearity. Thermistors and varistors are two such devices.

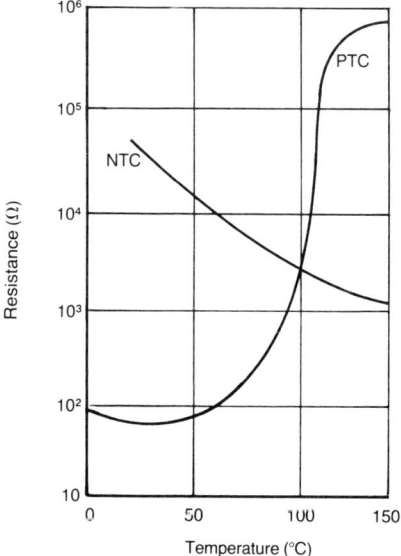

Figure 3.5 Resistance temperature characteristics of NTC and PTC thermistors

The thermistor (an acronym for *therm*al re*sistor*) as the name implies is a temperature-sensitive resistor. They are of two types. The negative temperature coefficient (NTC) type, whose resistance decreases with increase in temperature (Figure 3.5), are made from a sintered mixture of the oxides of nickel, zinc, copper and manganese. The opposite function is performed by the positive temperature coefficient (PTC) type, the resistance of which rises on heating. Thermistors are fabricated in discs, rods, beads and washers (Figure 3.6) covering resistances up 10^6 Ω and with a wide variety of temperature coefficients. They have a wide range of applications, including the control and measurement of temperature

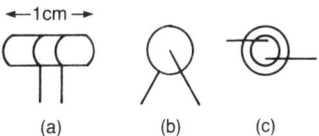

Figure 3.6. Thermistors. (a) Rod type; (b) disc type; (c) bead type

and the limiting of current or voltage. For example, the current flowing through a transistor junction is converted into heat which must be conducted away to prevent destruction of the junction. The collector current is controlled by the bias applied to the base of the transistor. An increase in temperature can increase the collector current beyond the safe limit. If the bias current is reduced, the collector current will be reduced and the transistor will return to its proper operating point. Hence increase in temperature can be reduced by controlling the bias to the transistor. By the use of a thermistor in the bias arrangement, compensation against the effects of increase in temperature can be achieved. In the circuit shown in

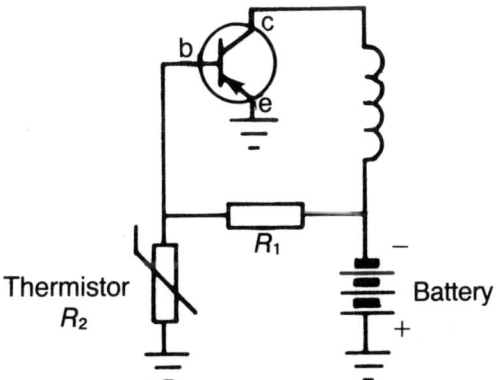

Figure 3.7. Simplified bias circuit for temperature-compensating thermistor

Figure 3.7 an NTC type thermistor is added to the bias network, the two resistors (*R1* and *R2*) make up a divider that supplies bias current to the base of the transistor. Any rise in temperature will lower the resistance of the thermistor and the voltage applied to the base of the transistor will decrease, the collector current remaining fairly constant.

In the case of the PTC type the increase in resistance is very steep within a comparatively narrow range of temperature. The PTC type is made of the ceramic semiconductor material barium titanate. This compound has a transiton temperature known as the Curie point at which the crystallites of barium titanate change from the tetragonal to the cubic structure. The transition is accompanied by a marked change in electrical properties, in particular the resistance which increases by several powers of 10 when the temperature reaches the Curie point (120 °C). The range of resistance increases can be selected by suitable doping of the barium titanate.

Varistors

A similar type of device is the varistor, a voltage-dependent resistor (VDR) whose resistance varies according to the applied voltage, but is independent of temperature. Consequently it functions as a voltage stabiliser and also to limit overvoltage. When the voltage across an ordinary resistor is doubled, the current also doubles (Ohm's Law) since the amount of resistance remains unchanged. When, however, the voltage across a varistor is doubled the current increases approximately 12 times showing that the amount of resistance decreases greatly when the voltage increases. It is made from a form of silicon carbide known as thyrite.

Capacitors

Together with the resistor the capacitor is the component most commonly encountered in electronic circuitry. A capacitor consists of two parallel metal conducting plates separated by an insulating (dielectric) material (Figure 3.8). When a voltage is applied to the plates equal and opposite electric charges appear on the plates and hence no net flow of electric charge takes place but only a displacement of charge, that is polarisation occurs. This fact implies capacitors are capable of storing charge and it is this property which is exploited. The storage capacity of a capacitor is termed the capacitance and is defined as the ratio of the charge to the voltage and is given by the following relationship.

$$C = \frac{Q}{V}$$

where C is the capacitance in farads (F), Q is the charge stored in the dielectric in coulombs (C) and V is the voltage across the capacitor.

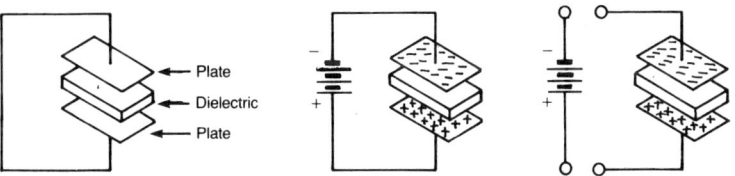

Figure 3.8. Diagram illustrating capacitor action. In (a) the plates are of the same potential and so there is no field across the capacitor. In (b) the capacitor is charged by using a battery, an electric field is produced resulting in an accumulation of negative charge on one plate and positive charge on the other. In (c) the battery is disconnected and the charges are trapped on the plates. Energy is actually stored on the electric field between the plates and is proportional to the applied voltage and the capacitance of the capacitor

When 1 C is stored in the dielectric with a potential difference of 1 V, the capacitance is 1 F. The farad, however, is far too large to be applied in practical electronic circuitry and it is usual to use the microfarad (μF) = 1×10^{-6} F and the picofarad (pF) = 1×10^{-12} F. For a given dielectric material, capacitance is increased by either increasing the surface area of the plates or decreasing the thickness and can be varied by varying the distance between the plates. These physical factors can be related to the capacitance by the following formula.

$$C(\text{pF}) = K \times \frac{A}{d} \times 0.08842$$

where A is the area (cms) of either plate, d is the distance between the plates and K is the dielectric constant (Table 3.2) of the material between the plates. The constant factor 0.08842 is the absolute permittivity of air in mks units. For example, if $A = 1$ m², $d = 1$ cm (or 10^{-2} m) and air is the dielectric

$$C = 1 \times \frac{1}{10^2} \times 8.85 \times 10^{-12} \text{ F}$$

$$= 100 \times 8.85 \times 10^{-12} \text{ F}$$

$$= 885 \times 10^{-12} \text{ F}$$

$$= 885 \text{ pF}$$

Capacitors are used for storing energy in the form of an electric field, producing an a.c. voltage drop and passing high frequencies while rejecting low frequencies. No current flows through a capacitor when subjected to direct current but when subjected to an a.c. voltage source it is alternately charged and discharged each half cycle. In this respect it differs from an inductor which is used to store energy in the form of an electromagnetic field producing an a.c. voltage drop, passing low frequencies while rejecting high frequencies.

The discrimination between a.c. and d.c. and the dependence of the a.c. capacitor on frequency makes the capacitor important in

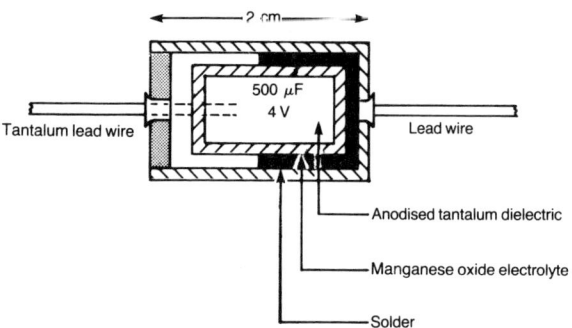

Figure 3.9. Construction of a tantalum electrolytic capacitor. The capacitance and voltage are usually marked on the body of the capacitor

radio, television and communication circuits.

In addition to the fixed type of capacitor (Figures 3.9 and 3.10) there exists a continuously variable capacitor (Figure 3.11) usually with air as the dielectric. Fixed metal vanes are connected together to

Capacitors   45

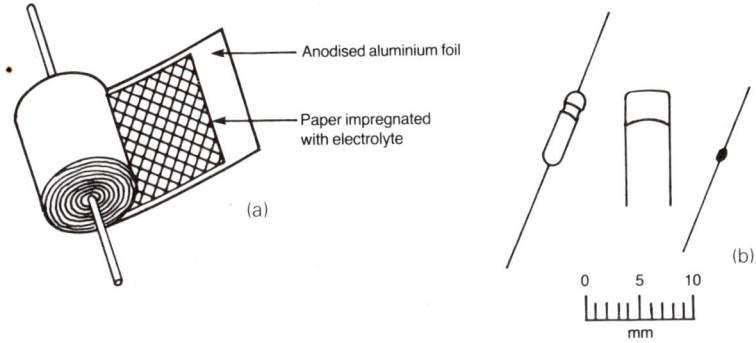

Figure 3.10. (a) Aluminium electrolytic capacitor. This is made by interleaving sheets of anodised aluminium foil between sheets of paper soaked in a suitable electrolyte. Leads are attached to the foil and the whole is rolled up into the familiar cylinder. (b) A selection of capacitors including a small capacitor used in hearing aids

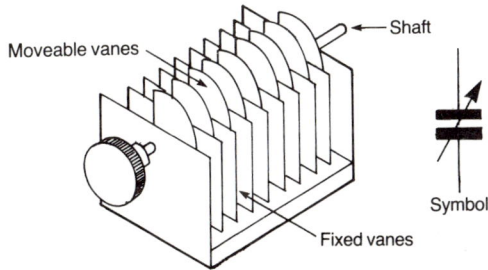

Figure 3.11. Variable air capacitor

form the stator. Movable vanes are connected together on a shaft and form the rotor. The capacitance is varied by rotating the shaft to make the rotor vanes mesh with the stator vanes thus creating different areas of overlap. Full mesh achieves maximum capacitance whereas moving the rotor completely out of mesh provides minimum capacitance. This type of capacitor is commonly used in radio sets as the means of tuning to different programmes.

Capacitors vary greatly according to size, application and the environmental conditions under which they are used. The main variable is the dielectric which stores the charge; the types of plates also differ depending on application.

The most important criterion used when choosing a dielectric is the relative permittivity K. This quantity, which in electrical engineering terms is the ratio of the absolute permittivity of the material E to the permittivity of free space, E_0, is a way of expressing how much charge a dielectric can store. Hence, if a vacuum dielectric between two plates of unit area can store unit charge, a mica dielectric with relative permittivity of 7 will store seven times as much. Some typical values of relative permittivity are given in Table 3.2.

Table 3.2 Relative permittivity (K) of dielectrics

Vacuum	1	Rubber	3
Air	1.00059	Polystyrene	2.5
Glass	4–6	Polyethylene terphthalate	3.1
Paper (dry)	3.5	Tantalum pentoxide	27.0
Mica	7	Barium titanate	1000–20 000
Aluminium Oxide	8.0	Porcelain	6.5

Other physical properties which are important when choosing a dielectric are the corona inception voltage and the coefficient of change of capacitance with temperature. Mechanical strength and dissipation factor are also of some importance.

Capacitors used in the electronics industries today fall into four main groups. These are electrolytic capacitors, paper capacitors, plastic capacitors and ceramic capacitors. The characteristics and applications of these types are set out in Table 3.3.

Electrolytic capacitors have a much smaller volume than the equivalent paper capacitors due to the extreme thinness (0.01–1 µm) of the dielectric layer between the plates. This layer is obtained by electrolytic oxidation of a suitable metal, usually a foil of high purity aluminium or tantalum (Figure 3.9). An electrolyte which is frequently used is water or glycol containing a mixture of boric acid and sodium borate or ammonium borate. Usually the metal electrode foils are rolled into a cylindrical shape with paper spacers to absorb the electrolyte being interspersed between them (Figure 3.10).

If breakdown occurs, the presence of the electrolyte ensures the self-healing of the dielectric, the oxide film reforming by oxidation of the underlying metal where the film was damaged. A development which has improved the frequency and temperature characteristics is

Table 3.3 Characteristics and applications of various types of capacitors

Type		Characteristics	Applications
Electrolytic	Aluminum	Polar, large capacitance-volume ratio. Limited temperature and frequency range	Smoothing, decoupling, by-pass, etc.
	Tantalum	Tantalum more expensive	Ta chosen for higher reliability and lower leakage current. Non-polar. Aluminium types suitable for intermittent use in a.c. circuits
Paper		Cheap, generally good performance, insulation resistance falls rapidly with increasing temperature	General purpose; power factor correction, blocking, audio and high frequency by-pass
Plastic	Polystyrene	High insulation resistance, low dielectric absorption and dissipation factor	Charge storage, filter circuits
	Polyethylene terephthalate	Capable of working at higher temperature than paper or polystyrene	Replacement for paper in higher temperature systems
Mica		High stability and low dissipation factor	Filter circuits. Standard capacitors
Ceramic	(1) Low permittivity	Low dissipation factor	May be used instead of mica in some circuits
	(2) Medium permittivity	Moderately high capacitance-volume ratios, controlled temperature coefficients of capacitance	LC resonance circuits
	(3) High permittivity	Large capacitance-volume ratio approaching that of electrolytics. Permittivity markedly voltage and temperature sensitive	By-pass, blocking and smoothing circuits
Evaporated silicone monoxide		Lower capacitance than tantalum types but superior dissipation factor	Microelectronic circuits

the substitution of the liquid by a semiconducting oxide such as manganese dioxide.

Unlike other types of capacitor the electrolytic capacitor is polarised, with the result that the applied voltage must always have the same polarity or the dielectric film will be destroyed.

Where a bipolar arrangement is required, two capacitors can be connected in series in such a way that the current is checked in both directions. A double capacitor of this type has a capacitance equal to that of only one of its component units, but is still considerably smaller than an equivalent paper capacitor.

It is possible to make a non-polar electrolytic capacitor, that is a capacitor that can be connected up with either polarity by using a special electrolyte which has a low leakage current and is self-healing for both directions of the voltage. Nitrates are suitable for the self-healing action in one direction and ions of magnesium, calcium, barium and aluminium will give self-healing in the other direction.

Paper capacitors are made from long strips of thin aluminium foil spaced by a paper dielectric, the whole being wound into a tight roll and sealed in metal or plastics cans.

The paper is usually impregnated with a liquid such as chlorinated-diphenyl, mineral oil or castor oil which seals the voids present in the paper and thus raises the safe working voltage. Another type of paper capacitor use metallised paper electrodes in which the metal foils are replaced by a metal sprayed on each side of the paper.

The wound unit has twice the capacitance it would have if the roll were unwound and laid out flat so as to form a parallel plate capacitor.

Paper capacitors are capable of being wound over a wide range of sizes and paper dielectric thickness and combined in series or parallel within a capacitor to supply a wide range of capacitance values and working voltages.

Because of their mechanical advantages, useful electrical characteristics and cheapness, impregnated paper dielectrics are used in applications such as power factor correction, filtering the ripple from the d.c. output of a rectifier, blocking, audio and high frequency by-bass.

Impurities contribute to dielectric loss and to d.c. leakage current. Impurity inclusions which would be innocuous under low voltage and low temperature may, at high temperature and high voltage, give rise to short circuits. For this reason paper capacitors are usually constructed with several sheets of paper, the number being determined by the voltage requirements.

Plastic films

Polyesters are among the best organic dielectrics known and are characterised by an insulation resistance which is higher than any other dielectric. The principal plastics dielectric in use at the present time is polyethylene terephthalate but new polyesters such as polycarbonate are now coming into use. These latter have all the advantages of the former with the additional benefits of lower power factor losses, greater frequency and temperature stability.

A power factor of 0.01 at 1 kHz and 20 °C is about a maximum for most terephthalates, but tests have indicated that power factors lower than 0.005 can be attained with polycarbonate capacitors, with insulation resistance values higher than 5×10^5 MΩ. Capacitor construction is similar to that used for paper dielectrics, tubular wound units containing plastic films being interposed between thin metal foils. Polycarbonate is available in sheets as thin as 2 μm. Polyethylene terephthalate can be obtained in sheets 3 μm thick. These figures compare very favourably with 5 μm for paper capacitors.

For the lower voltage applications a process known as metallising can be used in the production of film. This is accomplished by evaporating aluminium in vacuum from a heated crucible, and cooling to condense the metal vapour on the plastics film surface. To ensure that there is no variation in the thickness of the metal layer the resistance of the metallised film is continuously measured as it passes between two rollers and a feedback control is provided for controlling the film thickness. Resistance values within the range 1.5 to 4 Ω square are the normal target.

The excellent dielectric properties of the polyesters coupled with low cost have placed them in a favourable competitive position with regard to paper capacitors.

It is improbable, however, that complete replacement of paper capacitors by plastics capacitors will occur in medium and high voltage a.c. applications because plastics capacitors are limited in a.c. voltage since the corona inception level is only about 250–300 V a.c. Further, unlike paper units, plastics films are not normally self-healing; high electric current surges which occur at conducting sites in the film can produce deposits of carbon which lead to the presence of permanent short circuits.

Ceramics

Where large powers are generated at radio frequency, ceramic dielectric capacitors fill the requirement for currents up to 150 A at voltages of up to 5 kV. Their capacitance values may range from a few pF to 15 μF with a frequency of operation to 50 MHz. Ceramic capacitors are in general either high permittivity or temperature stable. The high permittivity capacitiors are relatively small in size but have high loss tangents and poor temperature stability. The latter may be improved, however, by the addition of modifying agents during the course of manufacture. This type of ceramic can be used for coupling and decoupling where losses and poor stability are not detrimental. For tuned circuits and filter networks where low losses and temperature coefficiencies are essential it is necessary to use dielectric materials of medium permittivity selected to have low losses and good temperature stability.

Magnesium silicate possesses a low permittivity (5-7) and power factor and a positive temperature coefficient of capacitance.

Titanium dioxide is the base material on which a range possessing a higher permittivity (80–90) and power factor are founded. Complex titanates of barium, magnesium and calcium have been developed into materials possessing a high and variable power factor and permittivities well in excess of 1000.

Vacuum deposition

Thickness of the metal electrodes formed by vacuum deposition is in the sub-micron range the thickness of the electrodes being usually of the order of 100–400 nm. Such thicknesses are obtained by evaporating a film of aluminium on to a glass substrate under a vacuum of 10^{-5} mmHg. This forms the base electrode, evaporation taking place as the result of electron bombardment or from electrically heated refractory metal crucibles. This is followed by evaporation of the dielectric typically silicon monoxide and then by a further aluminium evaporation which provides the second electrode.

Capacitors with vacuum deposited dielectrics do not compete with conventional components being at present usefully employed only in such microelectronic applications.

The impetus behind the use of vacuum techniques for preparation of dielectric films has been the need for high quality capacitors in microelectronics applications. In reality it represents a further stage in the development of the concepts of proportioning the thickness of the dielectric to the low voltages of modern circuitry.

For materials with high melting points an alternative method of obtaining thin film microcircuit capacitors is that of sputtering. This term is applied to a process in which atoms are ejected from a metal cathode as a result of bombardment by positive gas ions usually in a glow discharge. When accomplished in an atmosphere of oxygen, metal is deposited as oxide. The method has the advantage that the high temperatures necessary for the melting and evaporation of metals with high melting points is avoided.

Inductance

When the magnetic flux in a conductor is changing, a current is induced therein which lasts as long as change continues. The magnitude of the induced e.m.f. is proportional to the rate of change of flux and is represented by the equation

$$E = \frac{d\Phi}{dt} \tag{1}$$

where Φ is the magnetic flux with respect to time t. The induced e.m.f. acts in a sense to oppose the change in flux. Where the changing magnetic flux is due to changes in current flowing in a conductor, a counter e.m.f. is induced in the circuit, known as back e.m.f. This phenomenon is known as self-inductance, and is the ratio of flux to current.

$$L = \frac{\Phi}{I} \tag{2}$$

L is the symbol for inductance. The unit of inductance is the henry defined by the letter H. A circuit is said to have an inductance of 1 H if it develops a back e.m.f. of 1 V when the current through it changes at the rate of 1 A s^{-1}. The henry is a somewhat large unit and hence for all practical purposes a much smaller unit is used, that is the millihenry (1 mH = 10^{-3} H) or microhenry (1 μH = 10^{-6} H).

Inductive reactance

The opposition to a.c. current resulting from the inductance of a coil can be expressed in a manner similar to the opposition offered by resistance. In the case of inductance the opposition is known as induction reactance, the symbol being X_L. Instead of being a constant as in resistance, inductive reactance is directly proportional to the inductance and also to the frequency of the alternating current.

The equation for inductive reactance is

$$X_L = 2\pi f_L$$

where X_L is in ohms, f is the frequency in Hz and L is the inductance in henrys, the factor 2π being necessary to equate the result in ohms.

Solenoids

Although inductance is present to a small extent in a straight wire, in order to concentrate the magnetic field, inductors are made by coiling the conductor (usually copper wire) in the form of a solenoid. The fields of the various turns add together in the centre of the coil and hence the total effect will be far greater than for a straight wire. The induction may be further increased by winding the coil round a magnetic core. The induction of a solenoid depends on the number of turns, the area of the coil and on the permeability of the core material on which the coil is wound. It can be found from the following equation

$$L = \frac{\mu N^2 A}{l} \tag{3}$$

where L = inductance (H)
N = number of turns of the coil

μ = permeability of the magnetic material (H m¹)
A = cross sectional area (m²)
l = length of coil (m)

Hence the value of the induction is proportional to the square of the number of turns of the coil. Double the number of turns and the induction will be quadrupled. Permeability (μ) of a material is a ratio of the magnetic flux density to the magnetising force producing it and hence has to be taken into account. For an air coil permeability is not a factor since $\mu = 1$.

Equation (3) for the induction of a solenoid is derived as follows. For a given magnetic circuit there is a constant proportionality between the magnetomotive force and the resulting magnetic flux. This constant is known as reluctance (R_m) and can be defined as

$$R_m = \frac{F}{\Phi} \tag{4}$$

where F is magnetomotive force in ampere-turns and Φ is the magnetic flux.

The term is derived from the fact that opposition or resistance is registered in a magnetic circuit to the establishment of magnetic flux. It can be regarded as the counterpart of resistance in an electric circuit.

The magnetomotive force is the magnetic analogue of e.m.f. and is the line integral of the magnetising force around a closed path in a magnetic field. It is the product of the number of turns in a coil and the current in amperes which flows through it.

Hence $F = NI$ ampere-turns.

Substituting in equation (4)

$$R_m = \frac{NI}{\Phi} \quad \text{and} \quad \Phi = \frac{NI}{R_m} \tag{5}$$

From equation (2)

$$L = \frac{\Phi N}{I} = \frac{NI}{R_m} \times \frac{N}{I} = \frac{N^2}{R_m} \tag{6}$$

Reluctance is related to permeability (μ) by the expression

$$R_m = \frac{l}{A\mu} \tag{7}$$

where l is the length of the coil and A is the cross-sectional area of the coil (m²).
Substituting in equation (6)

$$L = \frac{N^2}{\dfrac{l}{\mu A}} = \frac{N^2 \mu A}{l} \text{ H}$$

Inductors

An inductor is a device for storing energy in a magnetic field. It may be regarded as the magnetic counterpart of a capacitor which stores energy in an electric field. Inductors are wound in the form of a spiral or coil (Figure 3.12) for the purpose of increasing the

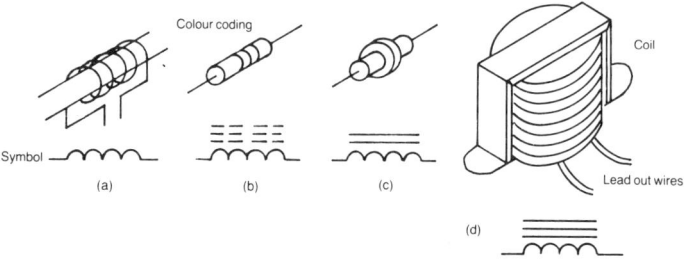

Figure 3.12. Inductor types. (a) Air coil; (b) r.f. inductor (ferrite core); (c) choke coil; (d) laminated iron-core inductor

inductance capacity. The method used in winding the coil, the number of turns, spacing of turns, permeability of the coil material and the physical shape of the coil (length, cross section etc.) all contribute to the specific electrical characteristics of the coil as does the presence or absence of a magnetic core. Inductors for frequencies of 1 kHz or less are usually required to be of a large inductance. To increase the inductance, it is necessary to decrease the reluctance and this involves the introduction of a ferromagnetic material within the magnetic field of the coil. For use at mains frequency the core is commonly fabricated from laminated sheets of iron or silicon steel, insulated from one another, the laminations reducing the undesirable eddy-current losses. By laminating the core, the eddy-current path is greatly increased in length and the resistance greatly increased. In applications where high power is not required, as for example in coupling transformers in audio frequency amplifiers, a core of high permeability material is used such as Permalloy (78% Ni, 22% Fe). Such an alloy has a permeability of 100 000 as compared with a value of 5000 for silicon steel.

For use in resonant circuits or transmission lines, inductors of high Q are required. This quality factor Q is the ratio of energy stored in an inductor during the time that the magnetic field is being established to the losses in the inductor during the same time. The Q factor is also called the 'figure of merit' of the inductor. Because of eddy-current losses it is not possible to use a laminated core, so a high resistance material such as ferrite or a core of powdered alloy is used.

Electrical resonance

As high permeability is undesirable, since as frequency increases so does eddy-current loss, it is usual to lower the permeability to about 100 by the inclusion of an air gap in the core. When both inductance and capacitance are present in a circuit, the capacitive effect dominates when the frequency of the supply current is low, while at high frequencies the reverse occurs and the circuit is inductive. If the inductive reactance should be equal to the capacitance, that is if the voltage across the capacitor is equal in magnitude but of opposite polarity to the voltage across the inductor then these opposing factors would cancel all reactance, the supply current being limited only by resistance. This condition is known as resonance and is important in radio, television and radar for the selection of a particular frequency from a complicated signal consisting of variations of voltage at different frequencies. It can discriminate against unwanted signals above and below the resonant frequency while at the same time affording a selective circuit for the desired signal. In addition, because both capacitors and inductors are available in variable forms such components can be tuned to achieve resonance for a particular signal frequency.

Where a large inductance is to be obtained in a small space multi-layer coils are used. Such coils are used in resonant circuits at broadcast and lower frequencies and as radio-frequency choke coils and as inductance standards for audio frequencies. A choke is used to signify an inductor that has a high impedance at the frequency of operation as compared with that of other circuit elements to which it may be connected. The name is derived from the fact that an inductance has a restricting or choking effect on an alternating current over and above its resistive effect.

Transformers

Inductance may be divided into two types (a) self-inductance and (b) mutual inductance. Self-inductance is a property of a single coil whereas mutual inductance requires two or more coils designed to have a mutual inductance between them. A component which utilises this form of inductance is the transformer. In essence it is a device without moving parts which transfers a.c. energy from one coil to the other through electromagnetic induction without an electrical connection between the two coils. One of the coils, the primary, receives power at a given voltage and frequency from the source; the other coil, the secondary, delivers power at a different voltage but the same frequency to the load (Figures 3.13 and 3.14).

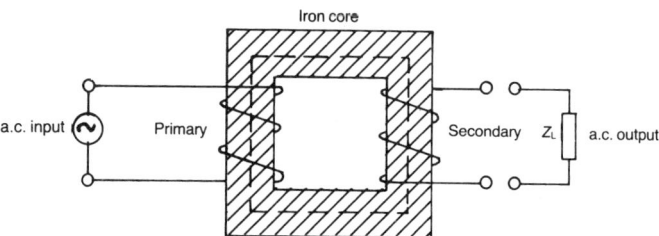

Figure 3.13. Iron-cored transformer

The transfer of energy from the primary to the secondary depends on the number of turns in each coil. The turns relationship is expressed by the formula

$$\frac{V_P}{V_S} = \frac{N_P}{N_S}$$

where V_P is the voltage across the primary winding, V_S is the voltage across the secondary winding, N_P is the number of turns in the primary winding and N_S is the number of turns in the secondary winding. By rearrangement of this equation the following

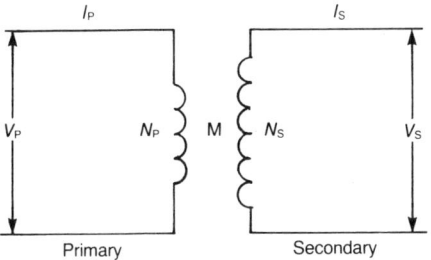

Figure 3.14. Circuit of a transformer

relationships are also valid

$$V_P N_S = N_P V_S; \quad V_P = \frac{N_P V_S}{N_S}; \quad V_S = \frac{V_P N_S}{N_P}; \quad V_S = \text{Turns ratio} \times V_P$$

Transfer of voltage from a high to a low voltage is achieved using a step-down transformer and the reverse that is transfer from a low to a high voltage using a step-up transformer. In Figure 3.15a the primary voltage of 200 V has been stepped down to 20 V whereas in Figure 3.15b the primary voltage has been stepped up to 2000 V. This ability to convert an a.c. voltage into a higher or lower output is one of the most important properties of a transformer.

The currents flowing in the primary and secondary follow a similar relationship but in the opposite sense

$$\frac{I_P}{I_S} = \frac{N_S}{N_P}$$

where I_P is the current flowing in the primary and I_S is the current flowing in the secondary winding. Hence a step-down in voltage produces a step-up in current and vice versa.

Autotransformer

This type of transformer differs from the normal type in that part of the winding is common to both primary and secondary the coil being tapped to give the desired voltage (Figure 3.16). The turns between the tap and one end constitutes one winding of the transformer and

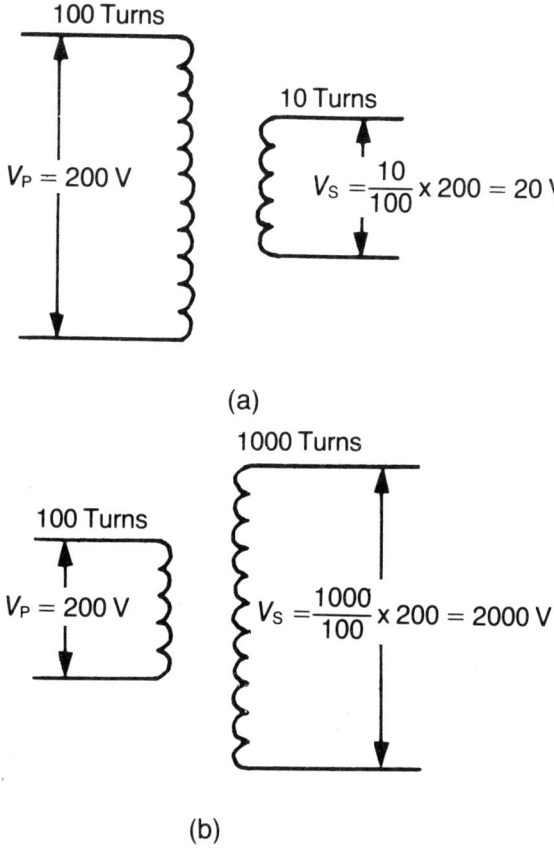

(a)

(b)

Figure 3.15. (a) Step-down transformer; (b) step-up transformer

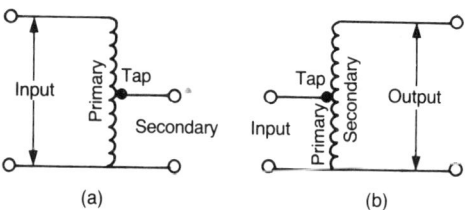

Figure 3.16. Autotransformers. (a) Step-down; (b) step-up

the full length of the coil constitutes the other winding.

In practice the transfer of energy from the primary to the

secondary winding is never 100%. The loss is occasioned by resistance of the coils, eddy currents arising through varying electromotive forces consequent on varying magnetic fields and from reactance caused by a leak of inductance and known as leakage or lost flux. To reduce this leakage the primary and secondary windings are wound on a core of ferromagnetic material of high permeability. This tends to reduce leakage by concentrating the lines of force and hence greatly increases the magnetic effect of the current. Ferromagnetic materials are composed of iron, nickel and cobalt. Transformers are used for impedance matching, for altering a.c. potentials and for electric isolation of d.c.

Impedance matching
When the source a.c. does not match the load impedance a loss of power is experienced. The power consumed by a load transducer (for example a loudspeaker) attached to the secondary winding is reflected back into the primary. As current flows through the secondary a magnetic field is set up which is reflected back into the primary. This cuts across the turns of the primary inducing a voltage which impedes the flow of current in the primary. The impedance of the load is in effect also reflected back into the primary. This impedance can be found from the formula

$$Z_P = Z_L \frac{(N_P)^2}{(N_S)} = a^2 Z_L$$

where Z_P = reflected impedance
Z_L = impedance of the load connected to the secondary
N_P = turns in the primary
N_S = turns in the secondary
$a = \frac{\text{primary turns}}{\text{secondary turns}}$ and is known as the transformer ratio.

Hence when a load is connected across the secondary a reduction of power from the maximum possible is experienced because a load is not matched to the source and part of the energy is transmitted (reflected) back to the source.

The turns ratio necessary to match the impedance can be found by adjustment of the above equation, viz

$$\frac{N_P}{N_S} = \frac{Z_P}{Z_S}$$

Example. A transistor audio amplifier requires a 2000 Ω load impedance. The loudspeaker has an impedance of 10 Ω. The primary has 800 turns. How many turns must the secondary winding have?
Answer

$$\frac{N_P}{N_S} = \frac{2000}{10} = \sqrt{200} = 14.1$$

$$N_S = \frac{N_P}{14.1} = \frac{800}{14.1} = 57 \text{ turns}$$

The transformer performs an important function in impedance matching for by adjusting a load impedance with a transformer maximum power is achieved, that is there is no reflection loss due to mismatch. By choosing the appropriate turns ratio impedances can be stepped up or stepped down. This can be a particularly important requirement in audio amplifier circuits, a transformer being used to match the transistor output impedance to the impedance of the load — the loudspeaker — maximum power thereby being provided to the loudspeaker when these two impedances are the same. A step-down transformer is needed to match the low impedance of the loudspeaker to the relatively high impedance of the transistor circuit.

4 Active electronic components I: semiconductor diodes

As outlined in chapter 1 the diode is a *p–n* junction non-linear device composed of an anode (*p*-type material) and a cathode (*n*-type) (Figure 4.1). The non-linearity arises from the fact that current through the device is not linearly proportional to the voltage across it, for it passes a greater current for one polarity than for the other. This unilateral conduction property is the basis for one of its chief functions namely power rectification. Inasmuch as almost all commercial electric power is alternating current, while all electronic circuits require direct current, there is a rectification circuit in nearly every piece of electronic equipment.

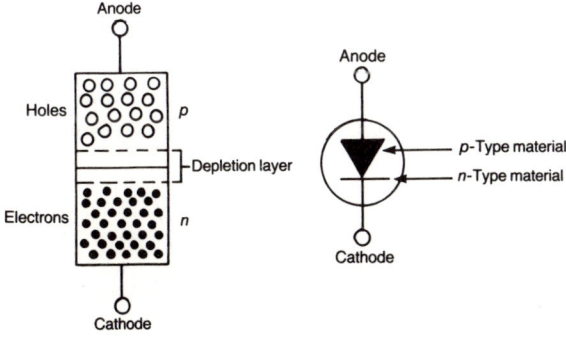

Figure 4.1. p–n junction diode with circuit symbol

Rectification

The simplest rectifier circuit (Figure 4.2) consists of a diode in series with a load, that is the device being fed from the power supply. The a.c. voltage source is represented by the sinusoidal waveform with

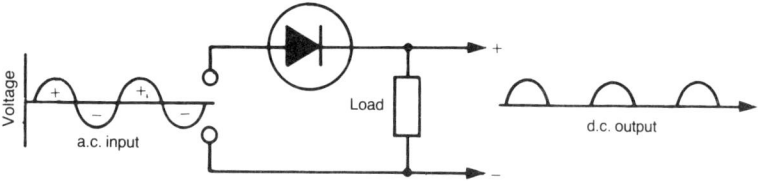

Figure 4.2. Basic diode rectifier circuit

alternate positive and negative cycles varying above and below zero potential. The effect of the alternating current is to apply forward and reverse bias to the diode. When the anode is forward biased, current flows through the junction and the load. Since the voltage drop across a forward-biased diode is relatively small, the current which flows during this part of the cycle is essentially the input voltage divided by the resistance of the load. On the other half of the cycle when the diode is reverse biased negligible current flows. The result is that the a.c. input is rectified by the p–n junction, the current flowing through the load resistor having the form of the top half of the sine wave.

Half-wave rectifier

The above is restricted to the description of a rectifier circuit in its simplest terms. For a more practical approach a somewhat more elaborate circuit is required (Figure 4.3). In order to step down the

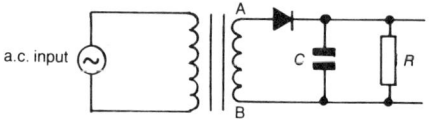

Figure 4.3. Half-wave rectifier circuit

240 V of the a.c. mains supply to a more suitable voltage required by

the electronic components, a transformer is necessary. It also serves to electrically isolate the mains supply from the output terminals necessary in the interests of safety. When point A is positive with respect to B, current will flow through the diode and the load R the result being a series of half cycles. When B is positive with respect to A reverse voltage is applied to the rectifier, no current flowing. Hence the diode passes current only during positive half cycles of the input so that the load voltage is a half-wave rectified current. The addition of a capacitor (C) serves to smooth the d.c. output thus providing a steady output. For obvious reasons this type of circuit is known as a half-wave rectification. It is not very satisfactory for since only half of each wave is utilised the efficiency is low. It is used only where small output current and power are required, for example battery charging.

Full-wave rectifier

Rectification in which current flows during both half cycles of the alternating current to produce unidirectional current is known as full-wave rectification. Two circuits commonly used to attain this end employ two diodes and four diodes, respectively, the latter being known as a bridge rectifier.

Double-diode rectifier
The rectifying circuit consists of a transformer, two junction diodes and a resistor (Figure 4.4). The centre tap of the transformer

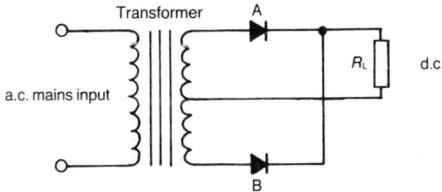

Figure 4.4. Two-diode full-wave rectifier

secondary is connected through the load resistance to the cathode sides of the diodes. During one half cycle of the a.c. input, the anode terminal of diode A is positive with respect to the cathode and the current passes through the diode and load resistance. During this time diode B is not conducting because the anode is negative with respect to the cathode. When the a.c. potential goes through zero the anode of diode A becomes negative and the diode stops conducting. The potential on the anode of B then becomes positive and diode B commences to conduct.

The effect of using two diodes is that the current flows alternately through both diodes, the first diode conducting for the positive half cycle, the second diode conducting for the negative half cycle. In this manner a continuous flow of direct current is attained. Although this circuit produces direct current it is unacceptable as a bias supply for many electronic devices because the voltage is pulsating, for it has alternating as well as direct components due to the rectified a.c. waveform.

This a.c. component in the output of a rectifier is known as a ripple current. These alternating components act as spurious signals and mask the desired ones. For instance in an audio amplifier such a power supply would produce a loud hum. To produce a steady d.c. voltage it is necessary to eliminate the pulsations and for this purpose filters are commonly used. Filters are widely used to select or reject voltages of a given frequency or range of frequencies. They are introduced here with reference to power supplies but they have many other important functions as well.

Filtering is frequently achieved by shunting the load with a capacitor. Its action depends upon the fact that the capacitor stores energy during the conduction period and delivers this energy to the load during the inverse or non-conducting period.

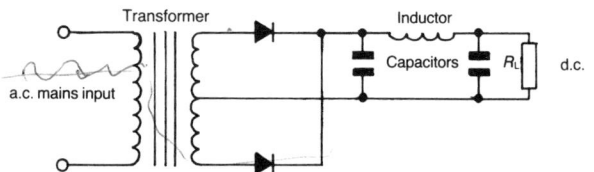

Figure 4.5. Rectifier circuit free from ripple

In this manner the time during which the current passes through the load is prolonged, the ripple considerably reduced. Inductors (chokes) are also used as filters and depend on their fundamental property to oppose any change in current. As a result any sudden

variations that occur in a circuit are smoothed out. The desirable feature of both is commonly combined in one circuit (Figure 4.5). The inductor offers a high series impedance to the harmonic terms and the capacitor offers a low shunt impedance to them. By using a filter that consists of two capacitors separated by an inductor a very smooth output may be obtained.

Bridge rectifier

This circuit uses four diodes in the form of a bridge (Figure 4.6), so named by analogy with the Wheatstone bridge. The alternating

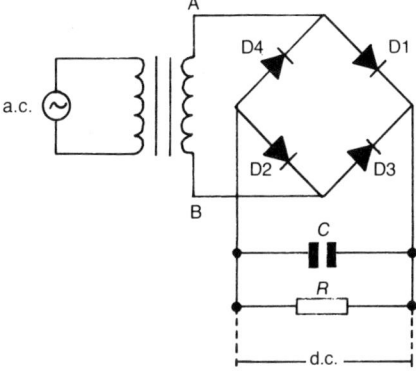

Figure 4.6. Full-wave bridge rectifier circuit

supply is connected across one diagonal and the d.c. output is taken from the other. When A is positive with respect to B, that is during the positive half cycle of the sine wave the diodes D1 and D2 are forward biased and are conducting since their p sides are more positive relative to the n sides. At the same time diodes D3 and D4 are reverse biased since their n sides experience negative voltage relative to their p sides.

During the negative half cycle of the sine wave the diodes D3 and D4 are forward biased and are conducting since their p sides are more positive relative to their n sides, while D1 and D2 are reverse biased since their n sides experience negative voltage relative to their p sides. In both cases current flows in R in the same direction. The action of the capacitor is similar to that explained in connection with the two diode rectifiers.

Active electronic components I: semiconductor diodes

In addition to its use as a rectifier the diode exhibits a number of other useful properties that are employed in many practical applications (Table 4.1). It may be used as a voltage regulator, a voltage variable capacitor, a switch, a light source and a means of converting light into electrical power and in fact it can be considered as a basic building block for many semiconductor devices. These related devices will now be considered.

Table 4.4 Semiconductor diodes

Type	Symbol	Material	Function
1. p–n junction diode		Si or Ge	Power conversion
2. Zener breakdown diode		Si	Constant voltage source used in voltage regulation
3. Avalanche (IMPATT) diodes		Si, GaAs	Microwave generators
4. Varactor		Si or GaAs	Voltage dependent capacitance, used in rf resonant circuits
5. Schottky-barrier		Si or GaAs	HF switching and detection
6. Tunnel diode		Ge or GaAs	Negative resistance device, used as HF oscillators
7. Solar cell		Si	Photovoltaic effects
8. Photodiodes		Ge, Se	Light sensors
9. Light-emitting diodes		GaAs	Display systems

Zener diode

By varying the doping density on each side of a semiconductor junction, diodes can be fabricated with specific breakdown voltages ranging from less than one volt to several hundred volts. The voltage at which this breakdown occurs when biased in the reverse direction is known as the Zener voltage named after Dr Carl Zener, an American physicist. It is to be noted that the phenomenon of breakdown as proposed by Zener is true of diodes operating only below 5 V (known as field effect diodes). Breakdown of the higher voltage types is due to a sudden build-up of current-avalanche multiplication, and hence they are known as avalanche diodes. This effect occurs in a semiconductor diode when the reverse voltage is high enough to accelerate the minority carriers so that they collide violently with the valence electrons of the atoms in the junction area. The collisions break some electrons loose which in turn become accelerated and break more valence electrons loose, this multiplication in effect creating an avalanche of electrons. The avalanche effect occurs only when the maximum reverse voltage is exceeded.

It is perhaps necessary to mention that at a critical value known as the Zener voltage the breakdown is usually non-destructive for when the overload is removed the diode returns to its original condition. In practice this usually involves limiting the current by means of a series resistor. Since the breakdown is non-destructive, diodes exhibiting this characteristics are exploited in voltage-reference devices and voltage-regulated power supplies and permit a convenient and simple establishment of accurate reference levels for instrumentation. An example of its application to voltage regulation is shown in (Figure 4.7). Assume this particular Zener diode has a

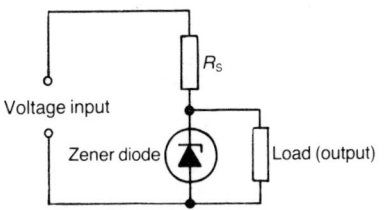

Figure 4.7. Zener diode voltage regulator circuit

breakdown voltage of 120 V, hence if any voltage in excess of 120 V is applied to it the diode will breakdown, causing a high reverse current through the diode and a constant 120 V drop across it. All voltages in excess of 120 V are dropped across series resistor R_s. As a consequence the voltage applied to the load will have a constant value of 120 V even though the input voltage is above 120 V.

Gunn diode

In recent years a new class of microwave devices has been developed which operate by the 'transfer electron' mechanism often referred to as the Gunn effect after J. B. Gunn who discovered the phenomenon in 1963 in gallium arsenide and indium phosphide. The Gunn effect involves the transfer of electrons from low-mass states in the conduction band of a suitable semiconductor to a high-mass state in fields of greater than 3 kV cm^{-1}. This gives rise to instabilities of the electron flow. A differential negative resistance (Figure 4.8) arises as

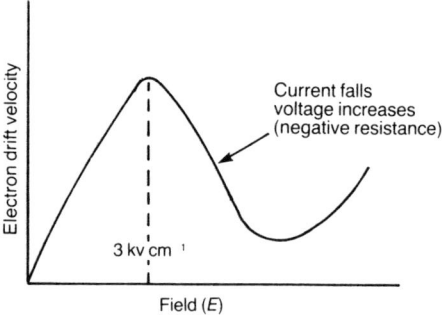

Figure 4.8. Velocity–field curve for gallium arsenide. At low fields the current increases linearly with voltage but above a threshold field of 3 kV cm^{-1} the electrons are transferred to the higher energy band, the drift velocity and overall current drop, the materials exhibiting what is known as a negative resistance effect. Termed the Gunn effect the negative resistance is used to convert energy from the electric field to produce oscillatory or amplification effects

a result of this transfer causing the carriers to nucleate into domains which travel across the device at the carrier drift velocity and disappear at the anode. As one high-field domain moves out another is created at the cathode and hence the frequency of oscillation is governed by the transit time across the device. In reality the Gunn effect is a high frequency instability that occurs in a suitable semiconductor when subjected to high electric fields. When a bar of such material is provided with non-rectifying (ohmic) contacts and then subjected to a high constant voltage it is found that the current in the bar is not constant but oscillates at a microwave frequency. When incorporated in a microwave circuit a Gunn diode when subjected to a r.f. signal superimposed on the d.c. voltage oscillates in various modes. A power output of 0.5 W and an efficiency of 6% have been obtained with diodes in continuous working operation and values of 7.5 W and 10% at 6 GHz with pulsed operation. Most of the electrical power supplied to the diodes is converted into heat (GaAs is a very poor heat conductor) and for c.w. operation of the diode this heat has to be dissipated.

The Gunn device (Figure 4.9) consists of an epitaxial layer of gallium arsenide grown on a substrate of strongly doped (tin) n-type GaAs mounted on a copper block for purposes of heat dissipation. A ceramic ring provides insulation between the two copper electrodes and a gold wire on the upper surface of the diode forms the electrical connection.

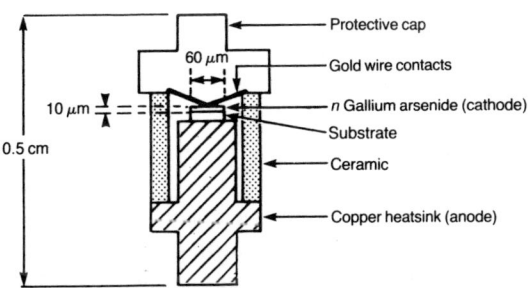

Figure 4.9. Basic construction of a Gunn diode

The Gunn effect can be used in various ways in the design of microwave oscillators. In the Gunn effect proper the transit time of the charge carriers through the semiconductor crystal plays an important part in determining the final operating frequency of the oscillation. However, in another form the transit time plays a minimal part and in fact the formation of domains is not involved for in the LSA (Limited Space change Accumulation) diode the frequency is so high that domains have insufficient time to form while the field is above threshold. This is an advantage in achieving conversion of power in the upper microwave range. LSA diodes can operate with an efficiency as high as 20% and at much higher power than the normal Gunn diode and also at very much higher frequencies. In general, oscillators built on these principles yield output power of 1 W at 10 GHz. Output power of hundreds of watts are possible in pulse operation round about 10 GHz and of 1 kW at 10 GHz. Output power of hundreds of watts are possible in pulse operation round about 10 GHz and of 1 kW at about 1 GHz. Combinations of a number of devices can be used to increase these levels.

Amplification

It has been found that n-type GaAs can also amplify signals in the vicinity of the transit-time frequency without oscillation if the product of the dopant impurity concentration (nd) and the diode length (l) is of the order of 10^{12} cm^{-2}. Above this limit domains can form and consequently oscillation will occur. Although the amplifiers are much less advanced than the Gunn oscillators they are already competitive to some extent with the low-power travelling-wave tubes.

Avalanche diode oscillators

The avalanche oscillator is another microwave source based on a negative resistance effect in semiconductors, both silicon and GaAs being used. These oscillators are sometimes called IMPATT (IMPact Avalanche and Transit Time) for they operate by a combination of the transit-time effect and the avalanche ionisation that occurs in the breakdown caused when a high reverse-bias voltage is applied to a p–n junction diode. In this condition the p–n junction exhibits negative resistance characteristics (due to the transferred-electron effect) in the microwave frequency range which can be exploited to generate microwave power and amplify microwave signals. IMPATT devices convert d.c. to microwave a.c.

signals and produce continuous powers ranging from 5 W at 12 GHz with an efficiency of 10% to 40 mW at 100 GHz with, however, a much lower efficiency. A further development with avalanche oscillators is the so-called TRAPATT (TRApped Plasma Avalanche Triggered Transit) mode of operation. Very high efficiencies (60%) have been obtained in pulse operation around a few gigahertz output with powers of 10 W. This high efficiency is the result of alternation between high voltage/low current and low voltage/high current conditions.

Microwave solid-state oscillators have become well established in new generation electronic equipment such as intruder alarms, man-pack radars, radar altimeters and distance measuring equipment. Gunn effect and avalanche diodes have played a major role in the development of these new systems and are replacing some of the older solid-state devices.

Varactor diode

The term varactor is a shortened term for variable reactor and refers to the voltage-variable capacitance of a reverse biased *p–n* junction. All diodes have a finite capacitance but a varactor is designed

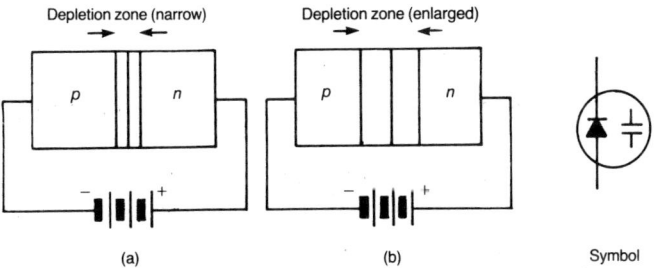

Figure 4.10. Varactor capacitance and reverse bias. (a) Low reverse bias; (b) high reverse bias. Capacitance of the diode can be varied by varying the width of the depletion zone which is controlled by the appalied bias. Increased reverse bias widens the depletion zone which increases the insulating volume and hence decreases the capacitance

specifically to exploit the capacitance present across the junction when a reverse bias is applied (Figure 4.10). As the latter is increased the depletion zone widens and since this zone constitutes the dielectric of the capacitance — two non-depleted *p* and *n* semiconductor regions forming the plates — it follows that the capacitance diminishes as the bias is increased. The fact that a reverse-biased *p–n* junction possesses a voltage-variable capacitance leads to several applications. For example, a varactor may be used in the tuning stage of a radio receiver to replace the bulky variable plate capacitor, the size of the resulting circuit thereby being greatly reduced. Other applications include use in harmonic generation, microwave frequency multiplication and active filters. The requirements for varactors for use in harmonic generators are based on optimising cut-off frequency and breakdown voltage. Gallium arsenide varactors are used extensively at output frequencies greater than 20 GHz where their higher cut-off frequencies dominate other factors. At low frequencies, however, their low breakdown value leads to reduced efficiency and in this connection the higher breakdown voltage of silicon junctions offers advantages.

Schottky-barrier diodes

This type of diode differs from the conventional semiconductor diode in that the junction consists of a metal and semiconductor (Figure 4.11). (The metals used to produce the barrier may consist of

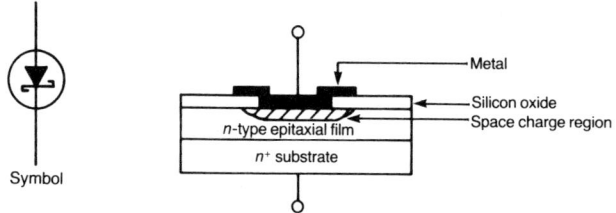

Figure 4.11. Schottky-barrier diode. This is a majority-carrier rectifier the rectifying junction being formed at the interface of a deposited metal layer and an n-type semiconductor

tin, aluminium, titanium or nickel.) Because of this modification the Schottky-barrier diode provides a better performance at higher frequencies for such microwave applications as detectors, rectifiers and switches. The reason for the better performance is that when the diode is forward biased electrons from the n-type silicon cross the junction into the metal and in this sense this is a majority carrier and hence current flow differs from that in conventional p–n junction diodes in that minority carriers do not take part in the process. On switching a p–n diode from 'on' to 'off' delay is experienced because the minority carriers stored at the junction must first be removed. The Schottky diode has negligible storage time because the current is carried by a majority carrier, in other words the metal-semiconductor junction has the effect of eliminating charge storage effects and enables very high switching speeds of less than 0.1 ns to be achieved.

Because of the greater uniformity the diode has low series resistance, a low noise figure and great resistance to pulse burn-out. It is widely used as the non-linear element in passive microwave frequency down converters. Another important application of the Schottky diode is as a component in integrated circuits.

Tunnel diode

This is a development of the ordinary p–n junction diode modified to

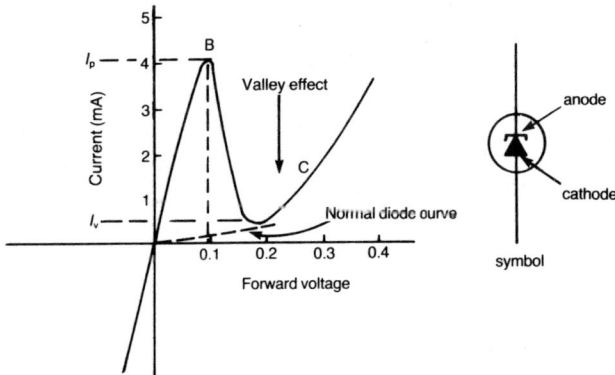

Figure 4.12. Current–voltage characteristic of a tunnel diode. At some critical voltage B the current decreases rapidly. This is the peak voltage at the point of maximum forward tunnelling. The area between B and C is known as the negative resistance area. The value of peak tunnelling current I_p and valley current I_v determines the magnitude of the negative resistance slope and hence the ratio I_p/I_v is used as a figure of merit

exhibit a negative resistance region in its current voltage characteristic. This modification takes the form of a very narrow depletion* (barrier) region obtained by heavy doping of both p- and n regions, typical doping being of the order of 10^{20} impurity atoms per cm^3 compared to 10^{13-15} impurity atoms for ordinary junction diodes. It will be noted as indicated in Figure 4.12 that as the voltage is increased the current increases much more rapidly than for a conventional diode. This steep rise is due to a phenomenon known as tunnelling in which the charge carriers move through the narrow $p-n$ junction instead of surmounting it. With increasing voltage a peak is reached and the with the voltage still increasing a fall commences thus exhibiting the negative resistance characteristic, a valley effect being created after which the current commences to rise again normally. The negative resistance characteristics of the tunnel diode together with its fast switching speeds, relative insensitivity to temperature, low power needs and small size and weight makes it commercially viable in video, microwave amplifying low power externally tuned oscillators, digital computer logic and low amplitude signal detectors. The tunnelling phase takes place at the velocity of light (3×10^8 ms^{-2}) but the switching speeds of a tunnel diode are reduce by the capacitance of the diode and also the external circuit to approximately 10^{12}s.

The diode has, however, the disadvantage that the voltage range over which it can be operated is limited to approximately 1 V. Further the diode is a two-terminal device and provides no isolation between the input and output circuit when used as either a shunt type switch or as a current amplifier. Many tunnel diode applications are being taken over by the Gunn diode because it can operate at higher frequencies and is cheaper to produce. It is to be noted that both diodes utilise negative resistance but different electron-behaviour mechanisms.

Unijunction transistor (UJT)

The term transistor applied to this device is really a misnomer for it has no amplifying power. As it has only a single $p-n$ junction and conducts more easily in one direction than in the other, it is more acurrately described as a double-based diode. It consists of a bar of

Unijunction transistor (UJT) 75

lightly-doped *n*-type silicon in which a heavily doped *p*-type emitter is formed on one side of the bar (Figure 4.13a).

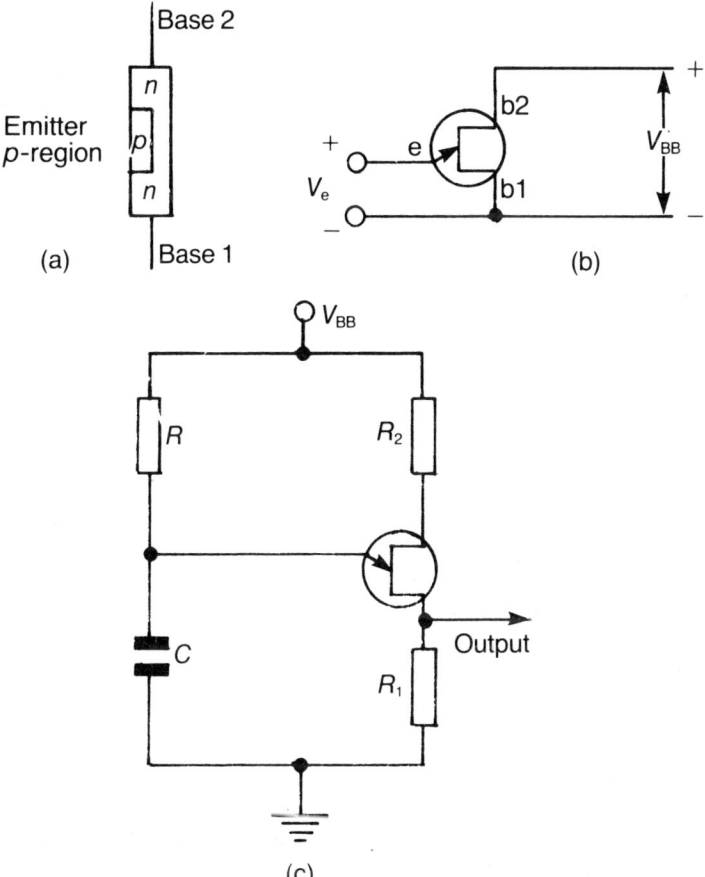

Figure 4.13. Unijunction transistor. (a) Structure; (b) symbol and circuit; (c) UJT trigger circuit

Ohmic connections are made to each end of the bar, these connections being known as base 1 and base 2. The emitter divides the bar into two resistances, in fact the bar acts as a potential divider. The ratio of the resistance between the emitter and base 1 to the total resistance of the bar is known as the intrinsic stand-off ratio, the ratio being denoted by the symbol η (eta).

When an input signal (V_E) applied to the emitter is less positive than the bias potential (ηV_{BB}) the junction is reverse biased and current flow is extremely small. Increasing the emitter voltage to a value greater than ηV_{BB} causes the junction to become forward biased and current carriers (holes) move into the bar which reduces the resistance and current flows freely. Hence the voltage from the emitter to base 1 controls the action of the UJT turning it off and on, and this triggering action is the basis of many thyristor circuits.

The emitter voltage at which triggering takes place is known as the peak-point voltage (V_p) and is given by the formula

$$V_p = V_D + \eta V_{BB}$$

when V_D is the forward voltage drop across the p–n junction and is of the order of about 0.6 V, the value of η lying within the range of 0.5 to 0.8 V. One form of UJT pulse trigger circuit is shown in Figure 4.13C. On the application of power the capacitor (C) is charged through resistor R from V_{BB} until the emitter voltage reaches the peak-point voltage. The UJT is then triggered into conduction, the capacitor discharging through R_1. When the emitter base voltage reaches a value of 1.5 V the UJT switches off and the capacitor commences recharging and the cycle is repeated. The time required for one cycle is of course very rapid, of the order of 10–15 μs duration.

Thyristors

The term thyristor is a generic name for a family of power control switches having three or more p–n junctions. The term is, however, more generally applied to the silicon controlled rectifier (SCR) and its modification the triac.

The SCR is a four-layered *pnpn* device (Figure 4.14) formed by diffusing an acceptor (gallim) into both surfaces of a basic *n*-type silicon wafer to obtain a p–n–p structure. A donor (phosphorus) is then diffused into part of one of the p-type regions so forming the p–n–p–n structure. Anode and cathode contacts are made to the top and bottom surfaces and a gate contact to the centre p-type region.

In practice the anode–cathode supply is often the a.c. mains, the thyristor being turned on periodically by short gate pulses and turned off by the reversal of the a.c. current.

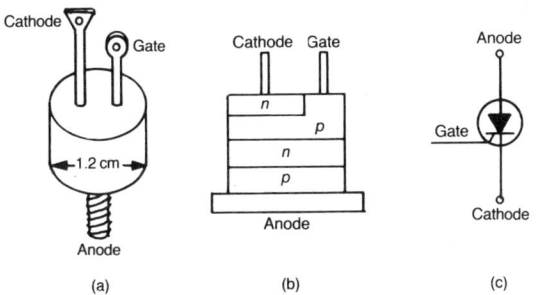

Figure 4.14. Silicon-controlled rectifier. (a) Device; (b) structure; (c) symbol

The SCR possesses the intrinsic advantages of semiconductor devices, that is good electrical efficiency, reliability and small physical size. It acts like a normal diode in the reverse direction, blocking current flow from cathode to anode. In the forward direction the thyristor is blocking until it is triggered or gated into conduction by a low-power pulse applied between the gate electrode and cathode. When conducting the device behaves like a closed siwtch, the carrier flow being limited only by the external circuit. If the current is reduced to zero the thyristor will revert to the blocking state until it is triggered again.

Voltage–current characteristic (Figure 4.15)

While the SCR is forward biased, in the absence of a gate signal, there is a small leakage current called the forward blocking current which remains small until a voltage known as the breakover voltage (V_B) is reached. At this point the current jumps rapidly to a high level, the anode voltage dropping to a low value. Once switched on the thyristor remains on provided the current remains greater than what is termed the holding current (I_H). The value of this current may be up to 50 mA.

When the voltage drops too low to sustain the holding current the device turns off. The value of the breakover voltage can be

78 Active electronic components I: semiconductor diodes

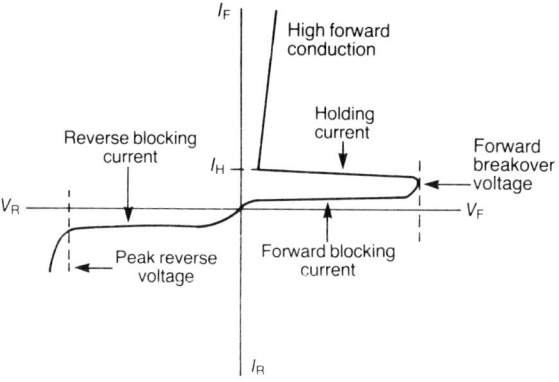

Figure 4.15. Voltage-current characteristic of a SCR

controlled by the level of gate current and provides a means whereby low levels of gate current can control high values of anode (load) current. SCRs are available with voltage ratings of 2000 V and current ratings of 1200 A. Thus control of 2 MW can be accomplished with a few watts of signal at the gate. This action is in contrast to the continuous control of current in a transistor the SCR action being that of a trigger, corresponding to the two states of an on-off switch. Figure 4.16 illustrates a simple switching current. As

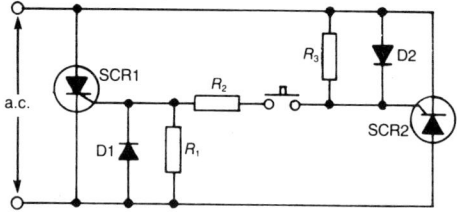

Figure 4.16. SCR switch circuit illustrating power control in an a.c. circuit

the thyristor is a unidirectional device, two SCRs are required to control alternating current. They are connected in parallel and in reverse configuration to each other, the anode of each being connected to the cathode of the other. They are therefore capable of conducting on alternate half cycles of the supply voltage. Diodes D1 and D2 conduct and supply current to the SCRs the resistors R_1 and R_2 limiting the gate current to a safe value.

SCR used as a rectifier

An example of this type of application is the use of a SCR to operate and control a d.c. motor or any other d.c. load from an a.c. supply (Figure 4.17). The SCR switches on when the forward breakover

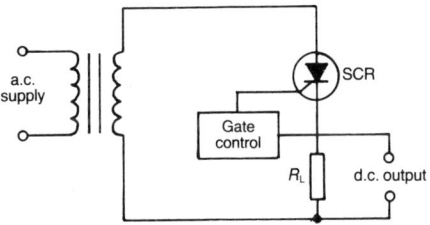

Figure 4.17. SCR rectifier circuit

voltage is reached on the postive alteration of the applied voltage. On the negative alternation it remains off, the a.c. supply voltage thus being rectified.

Trigger circuit

A pulse generation suitable for triggering a SCR frequently uses an unijunction transistor (Figure 4.18). Known as a relaxation

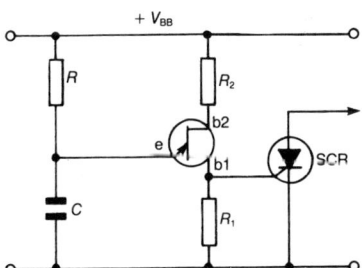

Figure 4.18. UJT relaxation pulse generator circuit

oscillator circuit the pulses are generated by charging the capacitor (C) through resistor (R) the emitter being reverse biased and hence cut off. However, once the peak voltage is reached the UJT is triggered into conduction and the capacitor rapidly discharges through R and triggers the SCR. Once discharged the charging cycle

Active electronic components I: semiconductor diodes

is recommenced, the continual charging and discharging of the capacitor giving rise to the so-called relaxation oscillations.

Triac

The circuit symbol is shown in Figure 4.19 from which it will be seen that the name derives from the three terminals, T1, T2 and a gate (TRI), the 'AC' indicating that the device controls alternating current. (A simplified cross section of the device is shown in Figure 4.19a.) It is a bidirectional thyristor which is capable of

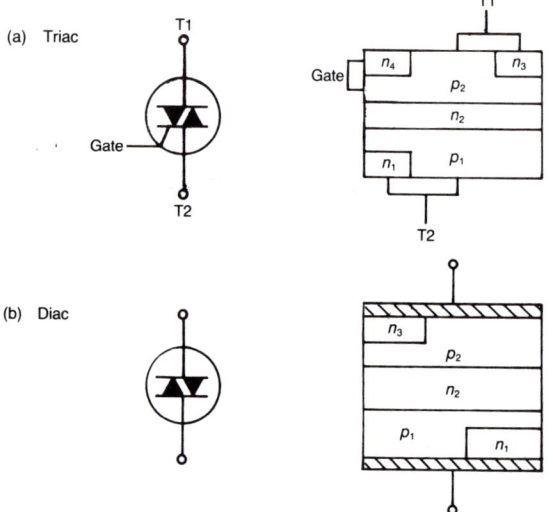

Figure 4.19. (a) Bidirectional thyristor or triac showing symbol and construction. (b) Diac, also a bidirectional device but without a gate designed for triggering

conducting or blocking current in either direction. It can be regarded as two *pnpn* devices connected in the inverse parallel configuration (that is in parallel with opposite orientations) but with a common gate electrode. Unlike the SCR the triac can be triggered with either a positive or negative pulse, and further since the SCR is a unidirectional device two are required to control an alternating current whereas the triac is capable of bidirectional switching.

Because the current can be conducted in either direction T1 and T2 are connected to both *p* and *n* regions. Similarly, because the device can be triggered by both negative and positive pulses the gate is likewise connected to both *p* and *n* regions. As with the SCR a specific voltage must be applied to initiate conduction. Since current flow is in either direction, the terminals are not designated as anode and cathode with the implication of polarity but are simply numbered.

Diac

The diac is a two terminal device (Figure 4.19) containing a dual arrangement of diodes side by side but having opposite polarities in their parallel configuration. The device is used as a special gating unit becuase of its characteristic of not conducting until a specific breakover voltage (40 V) is reached. When conduction does occur it will pass current in either direction. It is used primarily as a trigger unit in the gate circuits of triacs and SCRs. A simple form of trigger circuit for a triac using the diac is shown in Figure 4.20, in which the a.c. supply is connected in the circuit through the *RC* network. Capacitor *C* is charged during the positive half cycle of the supply through the variable resistor *VR*. When the voltage across *C* reaches

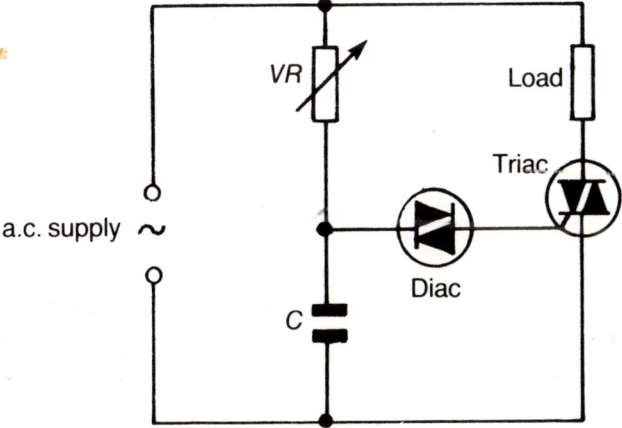

Figure 4.20. Trigger pulse generator circuit for a triac using a diac to provide the gating pulse

the breakover voltage of the diac, the diac changes to the conducting state and C discharges through the triac gate turning it on. During the following negative half cycle capacitor C is recharged, the diac breaks through again and the triac is retriggered. The charging rate of the capacitor can be varied by VR so that the voltage across the capacitor will reach the diac breakover voltage at different points in the half cycle and hence the power in the load may be continuously varied between zero and maximum.

Applications of thyristors and triacs

A convenient grouping of applications is into domestic and industrial, corresponding to light and heavy current uses. As regards the former the automatic washing machine is a good example. In this appliance the drum has to revolve at different speeds during the washing cycle. As the power in the load may be continuously varied by a SCR this forms a convenient means of controlling the current and hence the speed. Food mixers, sewing machines and electric drills also come into this category.

As regards industrial applications, the primary function of a thyristor is to modulate the power in a.c. and d.c. systems. The ability to switch and control power, leads to a wide range of applications. In particular the control of a.c. power is particularly useful since by triggering the SCR into the ON cycle at a variable time each cycle, proportional control over the average a.c. power in a load is achieved. Examples of this control are to be found in regulated power supplies, motor speed controls, temperature control of electric furnaces and ovens etc.

5 Active electronic components II: bipolar transistors

The bipolar transistor also known as the junction transistor is the basic functional element in electronic circuits. The property which makes them indispensable is their capacity for amplification of current, voltage and power. Bipolar means that carriers of both polarities — electrons and holes — are involved in the functioning of the device as opposed to field-effect transistors which involve carriers of only one type and are thus termed unipolar. The term junction refers to the fact that they consist basically of two p–n junctions formed within one continuous crystal as shown in Figure 5.1, in fact they may be considered as two series-connected back-to-

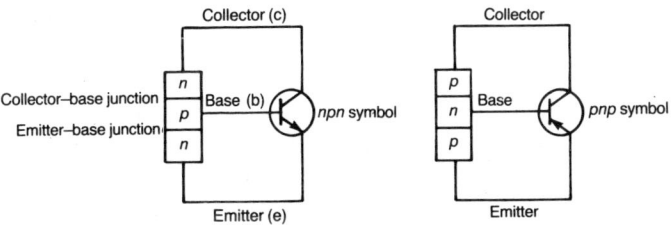

Figure 5.1. Diagrammatic representation of npn and pnp transistors

back diodes with a common base region. The two junctions give rise to three regions known as the emitter, base and collector, each region being provided with an ohmic contact to provide external connections.

The base region which lies between the emitter and collector controls the magnitude of the current flowing through the transistor and hence special consideration has to be given to its design. It has to be made very thin (typically 0.5 μm) in order that the charge carriers from the emitter pass straight through into the collector. A thin base also reduces the transit time for the injected carriers and hence increases the speed at which the device functions. It is only lightly doped and hence has low carrier concentration.

The emitter region injects charge carriers into the base and to this end is more heavily doped than the base. The charge carriers injected into the base diffuse across to the collector region and hence this action results in current flow in the external circuit. To minimise the adverse effect of junction capacitance it is only lightly doped.

A small combination of electrons and holes takes place in the base. If the base was thick nearly all the emitter electrons would combine with the holes in the base with the result that there would be a marked decrease in the number of electrons available to the collector current with a consequent large base–emitter current and a small base–collector current.

As the transistor is a two-junction, three-layer device it is apparent that there are two basic arrangements that can be used in its construction; either one in which *p*-material is located between two regions of *n*-type material or one in which *n*-material is located between two regions of *p*-type material. The first configuration gives rise to an *npn* transistor and the second to a *pnp* transistor. In the former the emitter arrow which indicates the direction of the conventional current points away from the transistor, whereas in the *pnp* the emitter arrow points into the transistor. The action of both is very similar, the essential difference being that in the *npn* type the charge carriers are electrons and in the *pnp* type the charge carriers are holes. Although they are both equally useful, as it is somewhat easier to mass produce *npn* devices, these tend to predominate in transistor circuits.

Transistor action

Figure 5.2a shows an *npn* transistor with the emitter connected to the negative terminal of one battery and the collector connected to the positive terminal of a second battery. The emitter–base junction is accordingly forward biased and the base–collector junction is reverse biased. The forward current at the emitter–base junction controls the current flow between the emitter and collector. Increase of forward bias increases the number of electrons injected into the

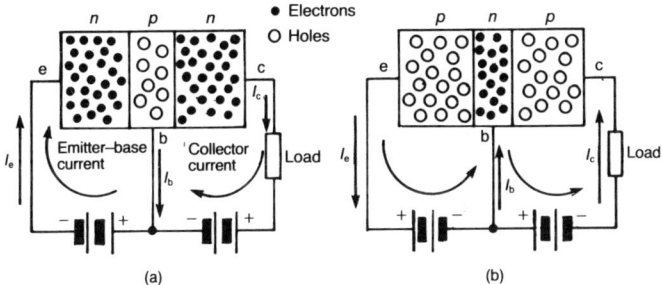

Figure 5.2. Current flow through an npn transistor (a) and through a pnp transistor (b). In the npn transistor the main current carriers are electrons, whereas in the pnp transistor they are holes

base and consequently increases the current flow from the emitter to the collector. Electrons which are the majority carriers are injected from the emitter into the base region. By the reverse biasing of the base–collector junction, the collector region is made positive with respect to the base and hence electrons which carry a negative charge penetrate into the base and flow across the collector junction and out through the load into the external current.

Referring back to the diode characteristic (p. 61) it was seen that only a fraction of a volt need be applied in the forward direction to start a current. Once the initial voltage (600 mV for a silicon diode) has been reached an extremely small change in voltage results in a large increase in current. Hence any increase in base voltage from 600 mV will produce an increase in emitter current. In fact an increase of only 20 mV will double the number of electrons entering the emitter and hence double the number of Carriers passing to the collector. Some of the electrons from the emitter will combine with the holes in the base region. Since there are relatively few mobile holes in the base region, the base current has a relatively low value. Hence current amplification has been achieved.

It is to be noted that the bias across the emitter and base is in the forward direction giving the junction between a low resistance. The bias across the base and collector is in the reverse direction giving the junction between a much higher resistance. Hence the term transistor which is derived from **trans**ferred **resistor**. Hence a suitable definition of a transistor is a three-terminal device in which the charge carriers within a semiconductor material flow from a *p–n* junction with a low input resistance to a *p–n* junction with a markedly higher output resistance. The bias of a transistor is the voltage applied to, or the current flowing between the emitter and base. This bias determines the operational characteristic of the transistor and can be considered as being either current bias or voltage bias or both. Current bias will vary from a few μA to a few hundred μA. The bias voltage will seldom exceed a maximum of 1 V.

Current and amplification

Amplification and gain are measures of the difference between the input and output. The input circuit of a transistor is associated with the input of carriers into the base region. The output circuit is associated with the flow of carriers from the emitter to the collector. The larger part of the current flow is from emitter to collector and

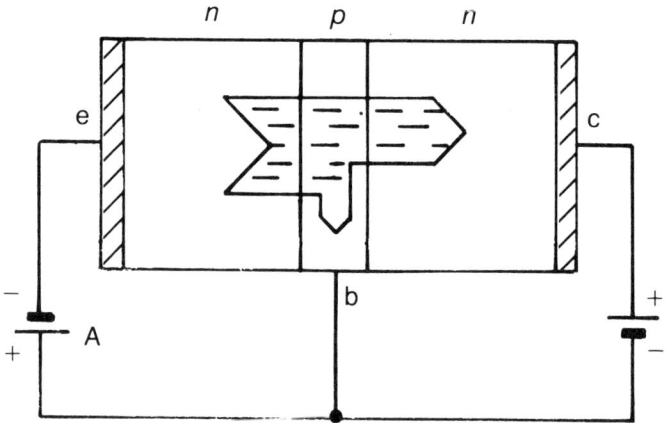

Figure 5.3. Direction of electron flow in an npn transistor. Electrons are repelled from the negative terminal of battery A and injected into the emitter junction. p-region (base) is only lightly doped (and hence highly resistant) and made very thin, and most of the electrons diffuse through the base and reach the emitter. A few electrons (about 2–5%) combine with holes in the p-region and are lost as charge carriers

only a very small current will flow between emitter and base (Figure 5.3).

Current gain of a transistor depends on (a) the number of electrons injected by the emitter and (b) the proportion of these that diffuse to the collector. The first of these factors is measured by the emitter efficiency which is defined as the ratio of the charge carriers injected into the base to the total emitter current. The second, known as the base transport factor, is the ratio of the carriers arriving at the collector to those leaving the emitter.

The emitter efficiency is largely determined by the conductivity ratio of the material on either side of the junction. To achieve high gain the conductivity and dimensions of the emitter region should be much greater than the corresponding values for the base region.

The base transport factor is a measure of the efficiency with which the carriers are transferred from the emitter to the collector. As the carriers diffuse through the base region some of them recombine (holes with electrons) forming the current which flows in the base region. As the base region is very thin and only lightly doped and the emitter region heavily doped it is possible for practically all the emitter current to flow into the collector.

Basic configurations

Since the transistor is a three-terminal device it follows that one terminal is always common to both input and output signals. Hence three basic amplifying circuits are possible (Figure 5.4). These are (a)

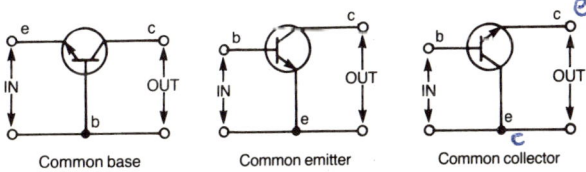

Figure 5.4. The three basic transistor amplifier configurations

common-emitter circuit with the input in the base-emitter circuit and the output in the collector-emitter circuit, the emitter being the element common to both the input and output circuits; (b) the common-base circuit with the input in the base-emitter circuit and the output in the base–collector circuit, the base being common to both the input and output circuits; (c) the common-collector circuit with the input to the base and output from the emitter circuit, the collector being the element common to both the input and output circuits.

Common-emitter circuit

This type of circuit is by far the most frequently used for it has the greatest power gain, substantial current and voltage gains and is especially advantageous in multistage application when a high gain is a primary requirement. A common-emitter amplifier stage with biasing from a single d.c. supply battery is shown in Figure 5.5. The

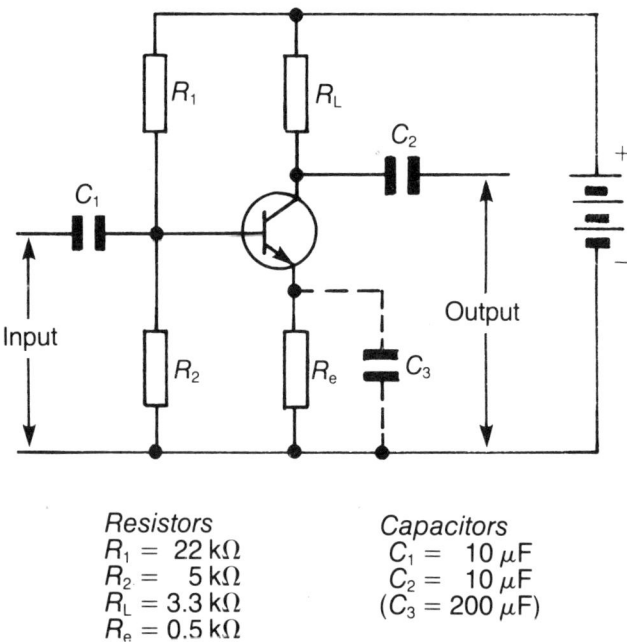

Resistors
$R_1 = 22\,\text{k}\Omega$
$R_2 = 5\,\text{k}\Omega$
$R_L = 3.3\,\text{k}\Omega$
$R_e = 0.5\,\text{k}\Omega$

Capacitors
$C_1 = 10\,\mu\text{F}$
$C_2 = 10\,\mu\text{F}$
$(C_3 = 200\,\mu\text{F})$

Figure 5.5. Common-emitter amplifier together with some typical component values

a.c. signal is applied between the base and the emitter and the output taken from the collector. For the transistor to operate the emitter junction must be forward biased, the resistors R_1 and R_2 setting the base voltage so that the emitter is forward biased while the collector junction is biased in the reverse direction. In series with the collector is the load resistor (R_L) the voltage developed by this being the output. The voltage gain of a transistor is largely determined by the value of this particular resistor since the voltage developed across it due to change in the collector current is far greater than that developed across the base resistor from the input signal.

Resistor R_e is included to minimise the effects of temperature changes in collector current. To prevent R_e from reducing the signal gain by current feedback a capacitor C_3 may be included in parallel with R_e.

The capacitors C_1 and C_2 are provided to prevent (block) the flow of direct current so that the d.c. bias conditions are in no wasy affected by the signal circuit. In this way the d.c. conditions at one stage are prevented from affecting the following stage, so that only d.c. signals are passed from one stage to the next one.

Transistor action

Current due to electrons enters the emitter region under the influence of the forward-biased base–emitter junction. The base region is very thin and the attraction force due to the reverse-biased collector junction making the collector region positive with respect to the base is very strong and hence most of the electrons jump the small potential barrier between emitter and base and pass through the base into the collector region. It is to be noted that not all the electrons pass through the base for some combine with holes in the base region and this gives rise to a small base current. Hence the collector current is always slightly less than the current flowing across the emitter junction.

An equation that defines the current action can be represented as follows, I_e being the emitter current

$$I_e = I_b + I_c$$

The input current is the base current I_b and the output current is the collector current I_c. Hence the equation is basic, relating the actual currents in a transistor. For a typical transistor with an emitter current of 1 mA the collector current might be 0.995 mA and the base current 0.005 mA (1 − 0.995). The ratio of the two currents I_c/I_b is the current gain and is signified by β (beta) or more correctly by h_{FE} (F for forward current gain and E for common emitter). Because the value of the base current is small in comparison with the collector current the value of h_{FE} can be very large. For example if I_e changes by 100 mA when I_b is changed by 0.2 mA the h_{FE} would be 100/0.2 or 500. Current gain in the common-emitter configuration ranges from 10 to 500 but multistage amplifiers may be constructed from any combination of the basic single-stage configuration to yield almost any gain. It is to be noted that due to deviations in the base–emitter voltage, spreads in h_{FE} can be very large and in fact can vary by a factor of three or more. Hence the current gain of an amplifier cannot be precisely known. When it is important to have a precise value of gain it is customary to use negative feedback (p. 00).

Voltage gain

The collector–base junction is reverse biased and consequently has a high impedance and hence a small change in the bias voltage produces a large change in collector voltage and a high voltage gain is obtained. Voltage gain is dependent upon the circuit current gain and the value of the load resistance and is given by

$$\frac{h_{FE} \times R_L}{R_I}$$

where h_{FE} = current gain
R_L = output load resistance
R_I = input resistance of the transistor between the emitter and base

Since R_I consists of the resistance of the forward-biased emitter junction its ohmic resistance is slight. On the other hand R_L, the load resistance, can be substantial and hence a large voltage and power gain is possible, for the input signal current flows practically unattenuated across an output resistance that is high compared with the input resistance. For instance, if the current gain is 30, load resistance 10 000 Ω and input resistance 1000 Ω then

$$\text{voltage gain} = \frac{30 \times 10\,000}{1000} = 300$$

Voltage divider

Resistors used in the manner illustrated in Figures 5.5 and 5.6, provide what is known as the voltage divider action. They are used to

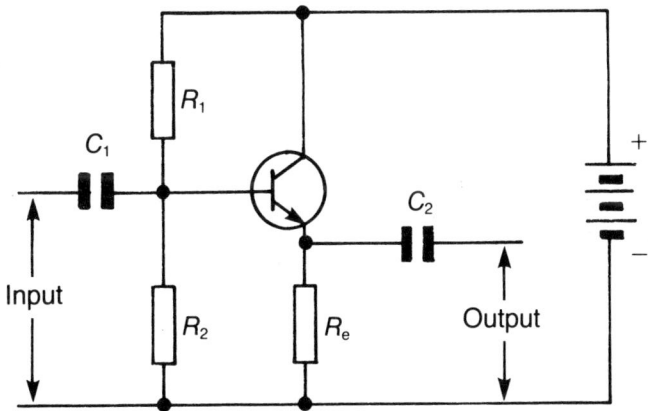

Figure 5.6. Common-collector amplifier

divide the voltage of the power supply into such values of potential as are required by the various parts of the circuit. It provides biased conditions for transistor amplifiers and in fact is the most important factor in applying bias voltage. In practically all transistor circuits the emitter–base junction must be forward biased to obtain electron flow through the transistor. For example in an *npn* transistor the base must be more positive (or less negative) than the emitter for current to flow from emitter to collector. The transistor is brought to the desired bias voltage by the voltage dividers R_1 and R_2 (Figure 5.5) which develop a particular value of positive voltage at the junction of R_1 and R_2 providing the desired biasing voltage for the base of the transistor.

Stabilisation

At high collector currents if precautions are not taken, the transistor will heat up, the increase in temperature increasing the collector current beyond the safe limit. This is known as thermal runaway. Precautions have therefore to be taken to check the effect which could eventually result in the destruction of the device. The increase in current flow that accompanies a temperature rise can be reduced by controlling the bias. Bias stabilisation can be applied by the introduction of an emitter resistor (R_e) which minimises the effect of temperature changes on collector current. Any increase in this current produces an increase in the emitter current and this causes a voltage drop across the emitter resistor. This reduces the base–emitter bias voltage which in turn leads to stabilisation. The emitter resistor also provides a degree of negative feedback which increases the bias stabilisation.

The voltage gain in a transistor is the ratio of the two resistors R_L and R_e and hence the value of R_e is dependent upon a compromise between stability and gain. An increase in the value of R_e in relation to R_L will increase stability but decrease the voltage gain. Similarly, the current gain of a transistor is the ratio of the two resistors R_2 and R_e and hence any increase in the value of R_e in relation to R_2 increases stability but decreased current gain. It is usual to make the value of R_e between 100 and 1000 Ω and not greater than one-fifth of R_L. As a basic rule R_2 should be about ten times the value of R_e.

In addition to the above arrangement a thermistor can be used for stabilisation purposes to compensate against the effect of increases in temperature. A thermistor is a device whose resistance decreases with increase in temperature (p. 40). When the temperature rises its resistance will fall as will the voltage across it so tending to maintain a constant base bias current.

Common-collector circuit

In this configuration the collector is the common point for the input and output circuits, the input signal being applied between the base and collector and taken off between the emitter and collector (Figure 5.6). The notable feature is the large input impedance virtually equal to that of the parallel circuit of R_1 and R_2. This is in contrast to the common-emitter circuit, where the input impedance is low because of the forward bias existing between the base and emitter. The output resistance is, however, low and hence it follows that the voltage gain is low, but a high current amplification can be obtained. The functions of the capacitors C_1 and C_2 are the same as for the common-emitter stage, as are the potential networks R_1 and R_2 which provide forward bias for the emitter–base junction. The chief advantage of the common-collector circuit is the readiness with which it may be direct coupled to any point in a circuit regardless of voltage. The circuit is often called the emitter-follower because the emitter voltage tends to follow the input voltage, the differences between the two being only the a.c. voltage across the base–emitter junction of the transistor which is quite small. Hence the output voltage must always be less than the input voltage and so the voltage gain is less than 1. The current gain is, however, high being approximately equal to that of the common-emitter circuit. The output resistance is very low (less than 100 Ω) since the emitter-to-collector resistance is low and there is no resistance in the collector circuit.

The external resistance of the collector circuit, that is the impedance presented by the transistor to the load is, however, very high (300 kΩ) and hence the emitter-follower circuit transforms a very high input impedance into a low output impedance; it is in fact an impedance transformer. Hence its main application is as a buffer i.e. an impedance matching device in which it can be connected between a high impedance source and a low impedance load without excessive loss of power due to unsuitable matching.

Common-base circuit

In this circuit the base is the common terminal between the emitter terminal and the collector terminal. The emitter current I_e is the input current and the collector current I_c is the output current (Figure 5.7). Since $I_e = I_b + I_c$ and since in this circuit I_e is greater than I_c by the value of I_b, the current gain I_c/I_e will always be slightly less than unity. Therefore there can be no current gain in a common-base circuit. However, because of the low impedance of the forward-biased emitter–base junction and the high impedance of the reverse-biased collector–base junction a sizable voltage gain is obtained. For instance, if we assume an input resistance of 200 Ω, a load resistance of 50 kΩ and a current gain of 0.98 the voltage gain is 0.98×50 k/200 = 245.

The common-base circuit is unsuitable for multistage amplification because its current and power gain are low when compared with the common-emitter. Also its low input impedance shunts the load resistance of any previous stage, thereby reducing the output voltage from that stage causing a corresponding fall in overall gain. However, its ability to operate at high frequencies makes it useful in v.h.f. and r.f. amplifiers. At such frequencies the

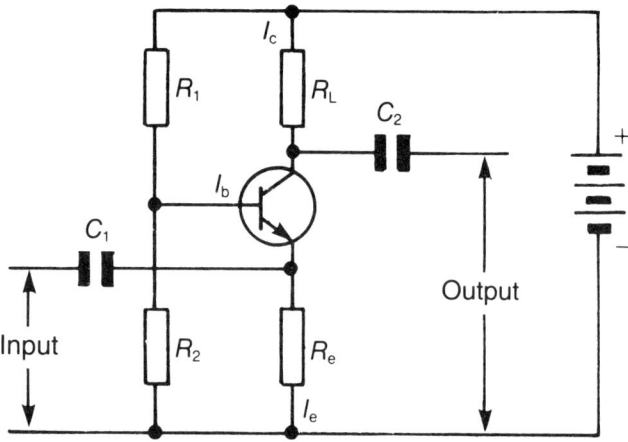

Figure 5.7. Common-base amplifier

circuit is more stable than common-emitter amplifiers because of the very small capacitance linking input and output circuits (the emitter–collector capacitance).

Summary

Approximate characteristics of the three basic circuit configurations are summarised in Table 5.1. Some representative types of transistors are shown in Figure 5.8.

Table 5.1 Characteristics of the three amplifier configurations

	Voltage gain	Current gain	Input impedance	Output impedance	Power gain
Common-emitter	High	High	Medium	High	High
Common-collector	Low	High	High	Low	Low
Common-base	High	Low	Low	High	Medium

Transistor characteristics

The behaviour of a transistor in a given configuration can be determined by a set of curves showing the relation between input and output currents and voltages. The graph so obtained relates the various characteristics and provides the necessary data as regards the operating current and voltages. Figure 5.9 shows a typical graph of the characteristics (*a-d*) in the common-emitter configuration.

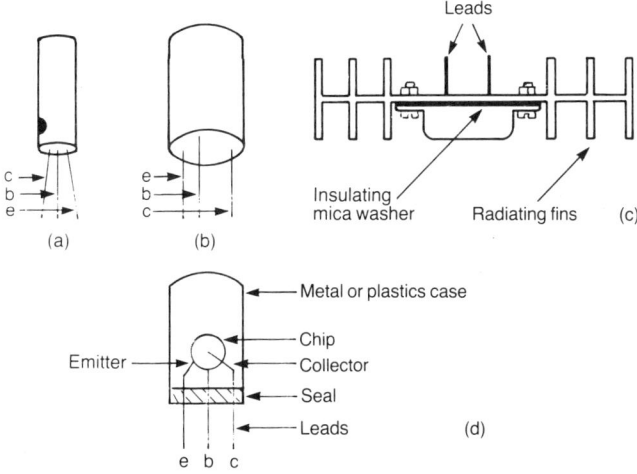

Figure 5.8. Common types of transistor. The transistor case is usually constructed of an opaque material since light will affect the operation. Lead identification: (a) a coloured dot or metal tag on one side usually identifies the collector; (b) wide spacing between the base and the collector identifies the latter. (c) In power transistors good heat dissipation is essential and a finned heat sink of a good heat conductor such as copper or aluminium is often used. (d) Cutaway view of a transistor

To illustrate the condition which occurs when the transistor is operating a load line is drawn, so called because the slope of the line depends on the value of the collector load resistance. Such a load line is represented by a diagonal line drawn through the characteristic curves. It shows how output signal current will change with input signal when a specified load resistance is used and indicates the collector voltage as a function of the collector current at a given voltage.

In the common-emitter circuit if the base voltage is zero there will be no base current and the collector voltage will be equal to the supply voltage V_{CC} (6 V). This is the point A on the load line the transistor is then said to be in the cut-off condition. If the base

96 Active electronic components II: bipolar transistors

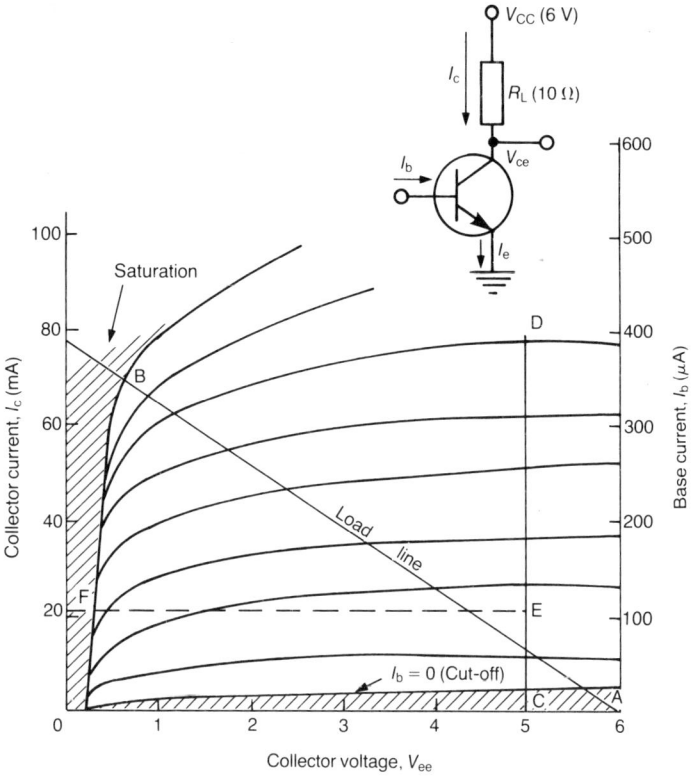

Figure 5.9. Transistor characteristics for the common-emitter configuration. Collector current I_c is plotted against collector voltage V_{ce} with base current I_b as a parameter. Using the load line for any variation in the input current I_b the corresponding values variations in I_c and V_{ce} may be ascertained

voltage is now increased, base current flows, producing a voltage drop across resistor R_L and so causing the output voltage at the collector to fall. Eventually a point will be reached when further increase in voltage will not result in any further increase in collector current. This is point B on the load line and the transistor is said to be in a saturated condition.

Intersection of the constant I_b curves with the load lines determines the operating points. For example:

If $I_b = 50\ \mu A$ then $I_c = 10\ mA$ $V_{ce} = 5.1\ V$
$I_b = 130\ \mu A$ $I_c = 27\ mA$ $V_{ce} = 4.0\ V$
$I_b = 180\ \mu A$ then $I_c = 38\ mA$ $V_{ce} = 3.4\ V$
$I_b = 350\ \mu A$ then $I_c = 70\ mA$ $V_{ce} = 0.5\ V$

Since the **horizontal** axes coincide the operating points are related. For instance a line CD drawn vertically from the point $V_{ce} = 5\ V$ represents the current variation I_c with a collector voltage of 5 V. With a base current of say 100 μA the operating point is found at E. The collector current at F is 20 mA giving a current amplification function for direct current of 200.

To determine the characteristics the circuit shown in Figure 5.10 is used. VR_1 is a variable resistor the purpose of which is to adjust the value of the base current, I_b, at a constant value, for example 200 μA

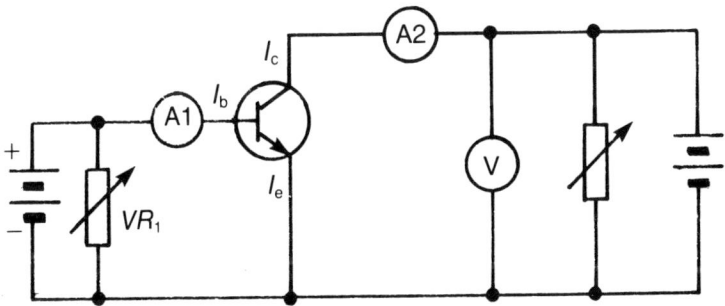

Figure 5.10. Circuit arrangement for resolution of characteristics for a common-emitter transistor

as indicated by the microammeter A1, the collector current I_c being noted on A2 for various value of collector voltages V_{ce} indicated on voltmeter V. The experiment is repeated for various values of the base current and the results plotted as shown in Figure 5.9.

Darlington pair

A dual transistor combination known as the Darlington pair is widely used in receivers and amplifiers. It is also known as the 'super alpha' pair because of the high current gain yielded by the circuit. It is essentially a method of cascading a pair of transistors (Figure 5.11), the distinguishing feature being that the two collectors are connected together, the emitter current of the first transistor forming the base current of the second transistor. The total current amplification will thus be practically equal to the product of the separate circuit gains of the two transistors.

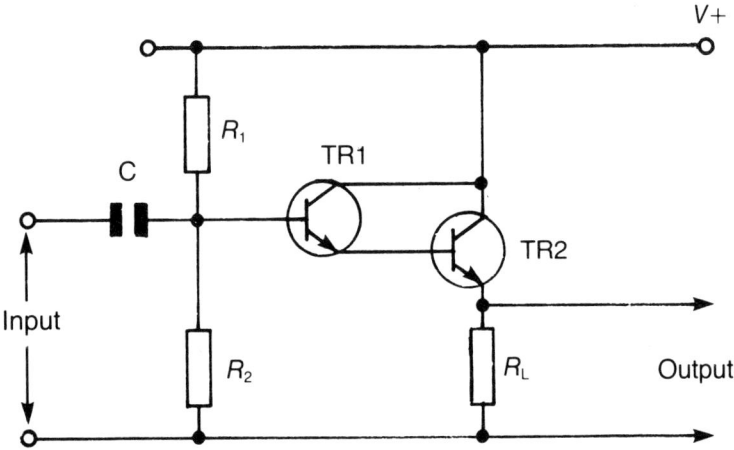

Figure 5.11. Darlington pair. Resistors R_1 and R_2 form a voltage divide for supplying bias to transister TR1. R_1 in series with the emitter of TR2 forms the output

The Darlington-pair configuration is very suitable for use in power amplifiers since only a relatively small driving power is required to obtain a large output current. Crossover distortion so troublesome with some types of amplifiers is much less prominent with this design.

6 Active components III: unipolar transistors

The conventional transistors so far considered consist basically of two *p–n* junctions in *npn* or *pnp* formation which depend for their operation on both majority and minority charge carriers. There is, however, a further type of transistor in which the operation depends entirely on majority carriers. Hence this type can be considered as unipolar for essentially only one kind of carrier, either holes or electrons, is concerned in their function. This second type of transistor is known as the field-effect transistor (FET). The term arises because amplification accrues when the current through two of its terminals is varied by an electric field arising from voltage applied to a third terminal.

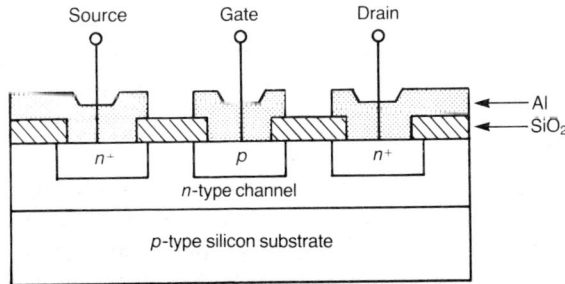

Figure 6.1. Cross section of a field-effect transistor (n-channel junction FET) showing source and drain connected by an n-type region which forms a conducting channel for the majority carriers. Current flow between the source and drain is controlled by a potential applied to the gate

FETs exhibit a variable resistance between two terminals, the source and the drain (Figure 6.1). The source is the electrode from which the majority carriers are emitted and flow towards the drain. The resistance is controlled by a potential applied to the third terminal, the gate, which controls the flow of majority carriers between the source and the drain. The conductive region through which the majority carriers flow is known as the channel.

Relative to bipolar transistors there are several significant differences and it is instructive at this stage to consider these.

(1) Current flow in FETs depends only on majority carriers.

(2) In an FET the current does not cross a junction in its path through the device whereas in bipolar transistors the main current crosses two junctions.

(3) The source, drain and gate fulfil the same function as the emitter, collector and base of a bipolar transistor, the essential difference being that in an FET the control element (the gate) is used to vary the resistance of the channel, whereas in a bipolar transistor the control element (the base) is used to control the current flowing across a forward-biased diode.

(4) FETs have a very high input resistance whereas for bipolar transistors input resistance is low.

(5) In FETs drain current decreases with increase in temperature and hence they are not prone to thermal runaway. In bipolar devices precautions have to be taken to guard against thermal effects.

(6) FETs are voltage-controlled devices whereas bipolar transistors are current controlled relying on small base-to-emitter current to control much larger amounts of collector-to-emitter current flow.

(7) In integrated circuits using bipolar transistors the individual circuit elements in a silicon chip have to be isolated from one another which involves a large number of separate steps during fabrication. No such isolation is required with FETs because the voltage applied to the source and drain ensure that no current can flow between them and the substrate. This not only makes the fabrication process simpler but also saves space on the chip, making for a much higher packing density.

Basic types

FETs are of two main types, the junction gate (JUGFET) and the insulated gate FET (IGFET), more commonly referred to as the metal oxide semiconductor transistor (MOST). They differ mainly in the manner in which the gate electrode is insulated from the channel. In the MOST the metal gate electrode is electrically insulated from the channel by a silicon oxide layer, this insulating layer being the reason why the device is known as the insulated gate FET. In the JFET the gate electrode is insulated from the channel not by an oxide layer but by a $p-n$ junction gate. A further difference is that while a direct connection is made to the gate of a JFET, there is only a capacitive connection to the gate of a MOST. The action of insulating layer representing the dielectric, the gate and channel forming the plates.

Symbols

FET circuit symbols (Figures 6.2 and 6.4) are indicative of polarity,

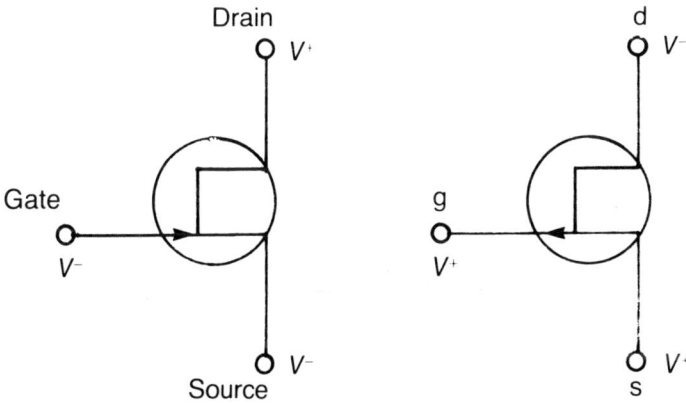

Figure 6.2. Symbols for the junction FET. Drain and source are linked by a solid line that symbolises the conducting channel between them. Direction of the arrowhead on the gate electrode indicates the channel material, pointing inwards for n-channel and outward for p-channel

gate construction (junction or insulated) and current mode (depletion or enhancement) the reason being that these three functions affect the electrical characteristics.

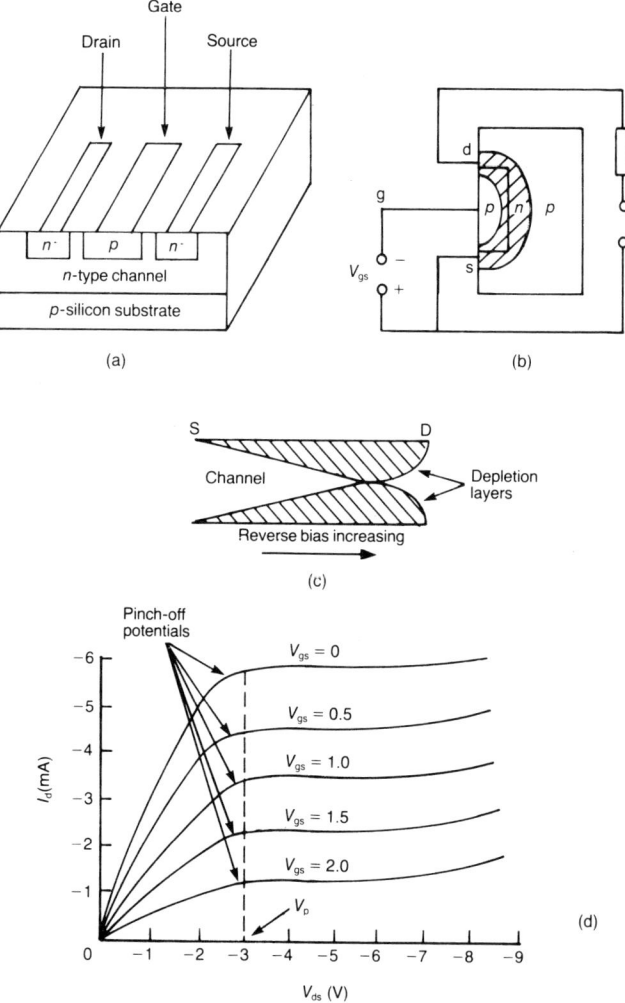

Figure 6.3. (a) Block outline of JUGFET. Cross sections (b) and (c) show the effect of bias on the depletion region. (d) Drain current (I_d) plotted against drain–source voltage (V_{ds}) for constant values of gate–source voltage (V_{gs})

Junction gate field-effect transistor

This consists of a lightly doped n-type epitaxial layer grown on a p-type substrate (Figure 6.1). Subsequently two heavily doped n-type regions (denoted by n$^+$) are added by diffusion. The n-type electrodes are the source and drain. Unlike the bipolar transistor the FET is normally operated in reverse bias by applying a negative voltage to the gate terminal. When an external supply (V_{ds}) is connected between the source and drain (Figure 6.3b) majority carriers flow through the channel. Current flow between source and drain is controlled by changes in the reverse bias of the gate-source and also in the resistance of the material between the two terminals. These two factors tend to reduce the conductivity of the channel and result in what is termed a depletion layer (Figure 6.3c). This leads to restriction of the cross-sectional area of the channel and hence increases the channel resistance with a consequent reduction of the current flow through it. At some critical voltage the depletion layer widens to such an extent that the conductivity path is reduced considerably. The voltage at which this occurs is called the pinch-off voltage. Under this condition the channel resistance increases to infinity and the conduction between source and drain is thus reduced to zero.

Referring to the drain characteristics graph in Figure 6.3d drain current (I_d) is plotted against drain-source voltage (V_{ds}) with constant values of gate-source voltage (V_{gs}). The curve indicates that for a given value of V_{ds} increase in the negative bias voltage decreases the drain current. This feature is referred to as the depletion mode of operation. The curves also indicate that the drain current rises at first in proportion to the increase in V_{ds}, until the latter reaches the pinch-off potential V_p from when it increases only fractionally. An estimate of V_p can be made from the value of V_{gs} that reduces I_d to approximately zero.

Applications

As a result of being voltage controlled, the reverse-biased junction gives rise to very high input impedance (5000-10 000 Ω) which is an outstanding feature of FETs. A high input impedance is desirable for many applications, for example in linear amplifiers used in digital integrated circuits, and for low frequency usage because a relatively low value of capacitance can be used to couple or decouple low-frequency voltages. Because of their extremely small physical construction they are used extensively as high-voltage amplifiers in colour television receivers and also as voltage controlled attenuators.

Metal oxide semiconductor transistor (MOST)

The MOST was an offshoot from the development of the planar technique in the early 1960s for making bipolar transistors. The doping of the silicon by diffusion makes use of the masking properties of a layer of silicon dioxide applied to the silicon substrate. It was found that this SiO_2 layer could act as an insulator between the semiconductor and the metal gate electrode of a FET,

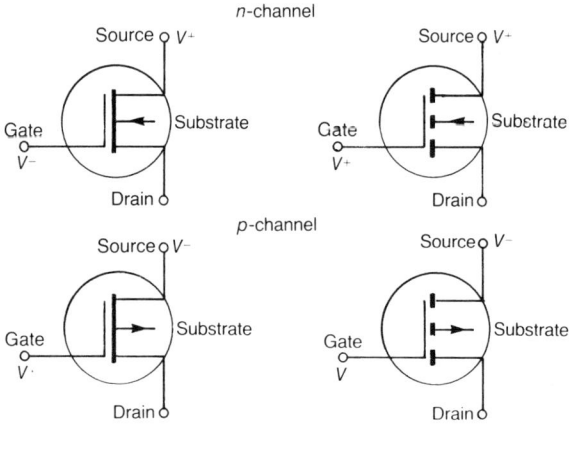

Figure 6.4. Insulated gate FETs (MOSTs) circuit symbols. The four possible types are shown; the difference between the depletion and enhancement devices is the vertical line representing the channel. For depletion the line is continuous signifying the continuous channel between source and drain; for enhancement devices the line is broken since the source and drain are separated. The difference between n-and p-channel devices is shown by the direction of the arrow on the substrate

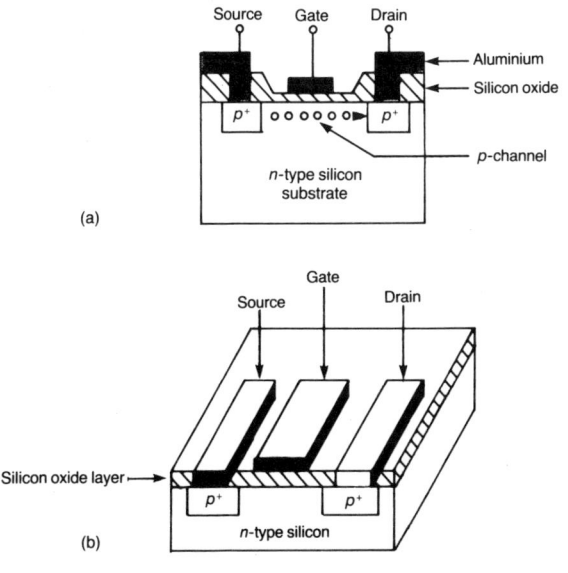

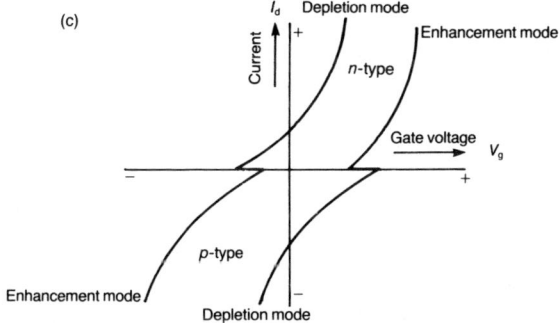

Figure 6.5. (a) Cross section of a p-channel enhancement mode MOST. Source and drain are formed from strongly doped p-type silicon. The metal gate is insulated from the substrate by a SiO_2 layer. (b) A block outline of the MOST. (c) MOST characteristics: in either the p- or n-type the threshold voltage may be either positive or negative so that there are four types of I_d-V_g characteristic

and this discovery resulted in the insulated gate FET or MOST. The MOST derives its name from its construction materials; metal, oxide and semiconductor. The metal, aluminium, serves as an electrical conductor, the oxide (silicon) acts as an electrical insulation at the surface of the device and the semiconductor is either n-type or p-type silicon. In the n-channel MOST the charge carriers are electrons whereas in the p-channel MOST the charge carriers are electrons there are four types of MOST — n-channel and p-channel types of both enhancement mode and depletion mode devices (Figure 6.4). The p-channel enhancement MOST is the most common type, a cross section of which is shown in Figure 6.5a. The device consists of two heavily doped p-regions, the source and drain, which have been diffused into a lightly doped n-type substrate. Typically the distance separating the source and drain is of the order of a few tenths of a mil. A thin insulating layer of SiO_2 is formed directly above the channel separating the source and drain. On top of this oxide insulation a layer of metal, usually aluminium, is deposited forming a third electrode, the gate.

The gate and channel constitute the two plates of a capacitor separated by the layer of SiO_2 which forms the dielectric. A charge on the gate induces an equal but opposite charge in the semiconductor channels. If the channel is p-type and a negative charge is applied to the gate holes are attracted into the channel and the resistance of the channel decreases. However, if a positive charge is applied to the gate some of the holes are repelled out of the channel and the resistance of the channel increases. Hence an electric field generated by a charge on the gate influences the motion of charge carriers in the channel.

Operations

When a negative bias is applied to the gate, the resulting electric field attracts positive holes, a conducting channel being formed between drain and source enabling current flow between the two regions (Figure 6.5a). If the negative gate voltage is increased, the depth of the channel increases enabling more current to flow. When current flow through a FET increases with increase in gate bias the effect is known as enhancement-mode operation because channel conduction is enhanced. On the other hand, when increased gate bias depletes the number of charge carriers in the channel thus reducing channel conductance, the arrangement is known as depletion-mode operation. These terms originally came from the use of the MOST as a switching device in digital circuits. Devices of the depletion type are normally open and require to be closed by a gate voltage which depletes the channel. Those of the enhancement type are normally closed and require to be opened by a gate voltage.

Threshold voltage

The minimum value of gate voltage just sufficient to cause channel formation is known as the gate threshold voltage (V_{th}). It is desirable in most cases to have an arrangement that produces a low V_{th}, since an IC with low V_{th} will operate at lower supply voltages than a high-threshold circuit. Typically V_{th} is between $+0.25$ V and $+1$ V for an n-channel MOST and between -2 V and -4 V for p-channel MOSTs. A further advantage is that low V_{th} is directly compatible with bipolar transistors and hence flexibility in ICs is possible. The switching speed of MOSTs is not particularly high because a comparatively high theshold voltage has to be applied before the device starts to conduct. Efforts to devise a technology that yields a lower threshold voltage has resulted in the silicon gate. In this approach the gate electrode is made not of aluminium but of polycrystalline silicon which, owing to the smaller parasitic capacitance, yields a lower threshold voltage and a higher switching speed. Parasitic capacitance between the source and drain regions can also be reduced by the silicon-on-sapphire (SOS) technique. In this case the starting material is a substrate of single crystal sapphire upon which a film of single crystal silicon is grown epitaxially. Further process steps are relatively standard with diffusion for the n- and p-regions. This technology yields a threefold improvement in speed–power over conventional MOST circuits.

Further improvements in the reduction of the threshold voltage at which the conducting channel forms have been attained by the incorporation of a layer of silicon nitride in the gate dielectric. This is accomplished by chemical vapour deposition. Silane gas (SiH_4) decomposes when it is heated releasing silicon and hydrogen. If a source of nitrogen such as ammonia is present during the decomposition of silane, silicon nitride is formed on the surface of the silicon dioxide wafer. The metal nitride oxide silicon transistor

(MNOST) also results in improved stability, characterised by the absence of volatility of stored information when used in computer memory devices. In bistable memory circuits, if the power supply fails information is lost. The MNOST in contrast is a non-volatile device.

MOSTs have a very high resistance at the gate by virtue of the insulating dielectric SiO_2 layer which isolates the gate from the substrate. Furthermore, the high resistance is obtained with either polarity of the gate–source voltage. Hence when a high value resistance is required in a circuit a MOST is often used. The replacement of another passive component with a MOST is also possible, for, by suitable modifications, the device can be made to function as a capacitor. Hence the MOST can be made to function as an amplifier, a resistor, a switch or a capacitor. As a result, when the MOST is used in ICs it is frequently the only component which appears in the circuit, that is it functions as all the required circuit components. For this reason MOST technology has come to dominate the manufacture of large-scale integrated circuits in which thousands of components are incorporated into a single device.

Complementary MOS transistors

An important trend in MOS transistor circuit arrangement is the advent of the complementary metal oxide semiconductor (CMOS) transistor. This arrangement uses MOS transistors having complementary symmetry, that is it is composed of both *n*- and *p*-channel MOS transistors. Both types are enhancement-mode devices, gate bias enhancing the conductivity between drain and source. Such circuits have significant advantages over conventional MOS circuits and bring to ICs the advantage of extremely low power requirement. With a 1.5-V supply, CMOS power dissipation is over two orders of magnitude less than that of the standard *p*-channel MOS and this extremely low power dissipation permits economical battery operation. This has led the way to the production of very small portable electronic devices such as handheld calculators and digital watches. This type of circuit is also faster than the MOS and more quiescent in operation. The advantages can be summed up as follows:

(1) Low power consumption.
(2) Good immunity to external noise.
(3) Insensitivity to power supply variations.
(4) Temperature-range capabilities −48 °C to 52 °C.

Areas of application

(1) Battery-operated equipment.
(2) Noisy environments.
(3) Aerospace systems.
(4) Digital communications equipment.
(5) Industrial electronics.

Characteristics

In outline a CMOS circuit chip has the cross section shown in Figure 6.6a from which it will be seen that the n-channel and p-channel MOSTs are connected together on a single substrate of n-type silicon. They are connected in series with their gates and drains connected together to form input and output terminals, respectively.

The reason for the low power dissipation can be discerned by considering the inverter circuit shown in Figure 6.6b. When the voltage at the input is zero, the gate–source voltage of the p-channel transistor is equal to the positive supply voltage V_{DD}. Since this is a p-channel device and the gate is negative with respect to the source the device is ON. When the input is V_{DD} the source–gate voltage of the p-channel transistor is zero so this transistor is OFF. However, at the n-channel transistor the voltage is V_{DD} but now with the gate positive relative to the source. Since this is an n-channel device it is turned on with the consequence that the output approaches zero. The point is that because of the complementary nature of the device one transistor is the complementary 'node' hence the p-channel does not conduct. The output voltage is then zero. If the voltage is now removed from the input, the p-channel transistor becomes conducting and the n-channel transistor is switched off. The output voltage is now V_{DD}. Power dissipation occurs only during a change of state. One transistor in the complementary 'node' is on and the other is off until a pulse arrives; then each of these changes state. The instant of change is the only time that significant current flows and hence power dissipation is kept extremely low, of the order of μA.

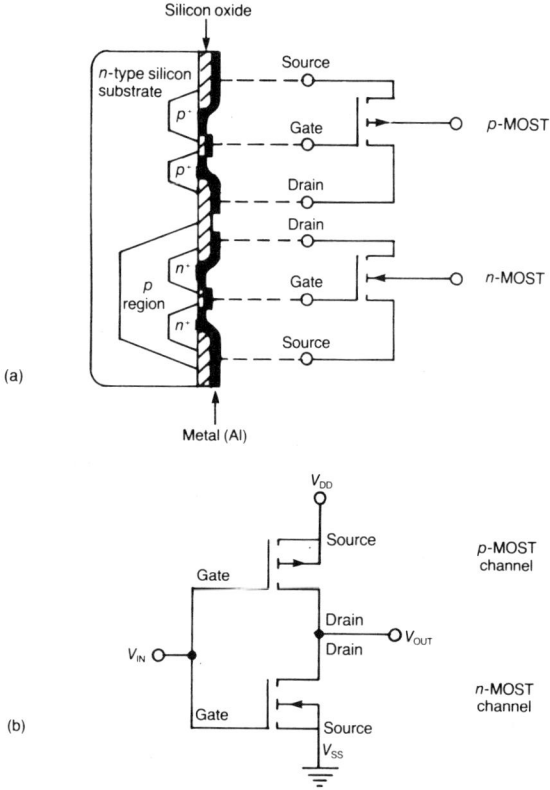

Figure 6.6. (a) The basic construction of CMOS. The solid line represents the metal which interconnects the devices. The n- and p-channel devices are connected in series. Only one of the two complementary devices is turned on at a time resulting in extremely low power dissipation. (b) Schematic diagram of CMOS inverter circuit. When the input is low the n-channel device is turned off, the gate of the p-channel device is turned on. Under these conditions the output goes high and is inverted with respect to the input. When the input goes high the n-channel device is turned on and the p-channel device is turned off resulting in a low output

The rest of the time the node is in the quiescent state — the driver transistor which is normally the n-channel device is on while the

p-channel load transistor is off. In this state the only power dissipation is caused by the leakage current of the switched off transistor. The dissipation is the product of supply voltage and leakage current and is of the order of nW per gate. On the other hand, in single channel MOS circuitry when a transistor is on, its load transistor is on too. There is therefore a significant power drain which in bipolar ICs ranges from 1 mW per gate for digital circuits to 80 mW for current mode logic.

Charge-coupled devices

The charge-coupled device (CCD) also known as a charge-transfer device (CTD) can be considered as an ingenious extension of the enhancement type MOS transistor. the micro-miniature nature of the structure and the ease of fabrication has resulted in the implementation of a device with an ultra-high-density electronic function in a very small area of a semiconductor chip. No solid-state impurity diffusion is necessary in its fabrication which is in contrast to the several diffusions required in the case of bipolar devices. This makes the surface area of the chip occupied by the CCD many times smaller than a bipolar device. Operation is very fast (100 MHz) and consumption of power very low (5 μW per bit); comparable figures for a silicon gate dynamic memory used in computers being about 10 MHz and 75 μW. These properties account for its importance in the construction of high-density memory arrays in computers and also in solid-state optical imaging devices for television cameras.

The CCD is constructed of metal gate control electrodes on a continuous SiO_2 dielectric layer grown on a single-crystal p- or n-type silicon substrate material (Figure 6.7). The device itself consists of a closely spaced array of MOS capacitors with a long row of gates located between a source and drain.

The three-layered structure creates and stores minority carriers in potential wells near the surface of the semiconductor. On the application of a more negative voltage to the adjacent electrode the minority carriers move from under one electrode to a closely adjacent electrode on the same substrate. Appropriate manipulation of potentials shifts the charge to the end of the device where a suitable output $p-n$ junction removes it.

To ensure correct operation of a CCD it is essential to maintain the silicon surface in depletion at all times. Application of a negative potential to the electrode has the effect of repelling the mobile

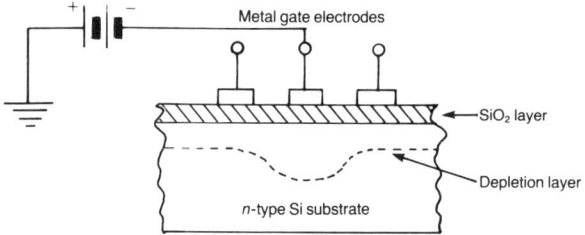

Figure 6.7. Schematic diagram of a CCD. A threshold voltage of about −1 to −2 V is applied to the substrate so that a uniform depletion layer forms beneath the electrodes

negatively charged majority carriers away from the underside of the silicon oxide. This results in the formation of a depletion region beneath the electrode, its extent being a function of the applied voltage. The negative potential at the gate attracts minority carriers (that is holes in an n-type substrate) into the depletion region until at the threshold voltage these minority carriers form a conducting channel at the oxide–semiconductor interface.

Charge coupling thus consists basically of a storage device in which information in the form of minority charge carriers is stored in specially defined areas under electrodes known as wells where depletion is momentarily deepened.

In operation a CCD has a threshold and negative bias voltage of about 1 or 2 V applied to the substrate so that a uniform depletion layer forms beneath the electrodes. Once a charge packet has been introduced to the CCD a more negative voltage, say 5 V, is applied to the second gate (Figure 6.8) creating a deeper depletion beneath it. When the voltage on the second gate is dropped and a voltage simultaneously applied to the third gate the charge packet stored under the second gate is attracted to and trapped under the third gate. The charge is thus transferred from beneath one electrode to the next along the silicon surface in clocked shift-register fashion by manipulation of the voltages on the control electrodes that constrain the charge. To ensure that the charge will not flow back the applied voltage on any two adjacent electrodes must always be lower than the electrode receiving the charge. Spacing between the electrodes has to be very small (2 μm) so that the charge is not lost on transferring from well to well by falling into 'traps' caused by the surface states or lost by diffusion.

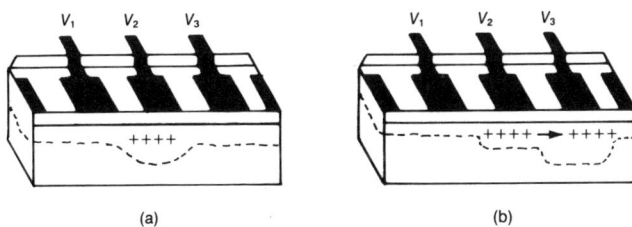

Figure 6.8. Section of a three-phase CCD. (a) A charge packet is stored in a potential well formed by a negative voltage V_2 greater than V_1. (b) Charge transfer takes place by applying a negative voltage V_3 greater than V_2. Following the same procedures the charge eventually reaches the output terminal

A shift register has been mentioned in the previous paragraph. The shift register is a very important device used in computers to store information temporarily and then to serialise the information. When a digital computer operates successively on each bit rather than sequentially, the procedure is referred to as serial operation. A leading example of an electronic memory device organised in this way is the CCD and the manner in which it operates is shown in

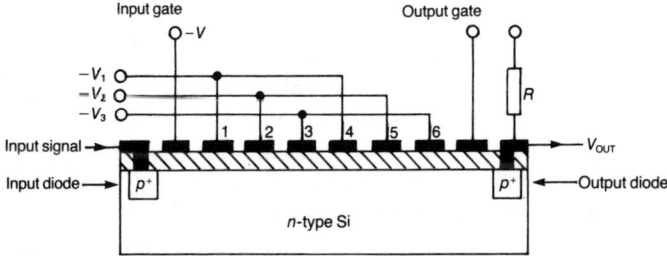

Figure 6.9. Longitudinal cross section of a CCD shift register

Figure 6.9. The electrodes are connected in groups of three and operated with a three-phase voltage supply. The left-hand column shows the potentials applied to each electrode, the potential V_1 being sufficient to provide a depletion region in the device. V_2 is larger than V_1 and produces a storage site. V_3 being still larger effects the transfer. Charges introduced into the device are initially stored under electrode 2. Potential on 3 is increased and the charge moves over one position to the right. Subsequently, the potentials on the electrodes are readjusted so that the quiescent storage sites are restored. In this manner charge packets are transferred from one electrode to the next and so to the output of the device. In digital application for CCDs a logic '0' state corresponds to the situation where zero channel charge is stored under a given electrode. Conversely a logic '1' state corresponds to the storage of a maximum charge.

Input circuit

Charge is injected into the device by the use of a p-region of an input diode which is an infinite source of holes. When a negative potential is applied to the input electrode it creates an inversion channel between the input diode and the first gate electrode. As a result the charge packet (that is holes) is conducted to the potential well under the first gate which is biased to attract the holes. Charge packets generated in this manner are transferred along the CCD as described previously.

Output circuit

This consists of a p–n junction diode with a large reverse bias positioned in such a manner that its depletion region couples with that of the final storage electrode. Because the reverse junction potential is more negative than the last gate potential, the charge present in this gate will be collected by the output diode, and appears as a voltage output developed across resistor R.

A memory device in which digits are stored as long strings of bits is fabricated by building a CCD with perhaps 1000 gates. Since the packing density can be very high many such devices can be fitted on to one chip. It has the advantage of using processing and lithographic techniques essentially the same as those employed for the fabrication of high mobility charge carriers.

Transfer efficiency

The effectiveness of the operation is measured by the efficiency of charge transfer and can be defined as the fraction of charge transferred when a charge packet moves from under one gate electrode to the next. The charge loss can be considered as being made up of two factors: (a) the fractional charge loss during the transfer across the gap between the electrodes (q/T) and (b) the fractional charge left behind under the elctrodes, the so-called residual charge (q/R) due to trapping surface states. If q is the charge which each packet carries and n is the transfer efficiency $n = (1 - qT - q/R)100$. The charge transfer efficiency has been found to be 98.5%. The fractional charge lost during transfer depends on:

(1) Width of gap between the transfer electrodes. If this gap can be kept below 2 μm and the electrode size below 10 μm operation becomes very efficient.
(2) Strength of input signal.
(3) Speed of transfer (frequency of operation).

Optoelectric application

Charge packets in the CCDs that have described so far are introduced electrically. Charge packets can also be generated optically for CCDs are excellent optical detectors throughout the visible spectrum. When one side of a CCD array is exposed to light, potential wells are created, packets of charge accumulate beneath the electrodes. With appropriate circuitry the charge can be transferred from one electrode to the next and out of the array a line at a time producing a scanning effect similar to that of a television camera tube. In operation, charges are introduced into the device when light from a scene is focused on to the surface of the device. The incident light creates hole–electron pairs in the substrate which under the potential beneath each storage electrode are collected as a charge packet. In this manner a spatial charge representation of the scene is stored in the device. It is transferred off the device when clock voltages are applied to the electrodes transferring each charge packet from storage site to storage site across the device until all charges reach an output diode.

116 Active components III: unipolar transistors

The technique can be applied to vidicons of all types, from computer terminal readouts to commercial television cameras. Unlike conventional television camera tubes, CCDs do not require the complex high-voltage vacuum technology of a scanning electron beam.

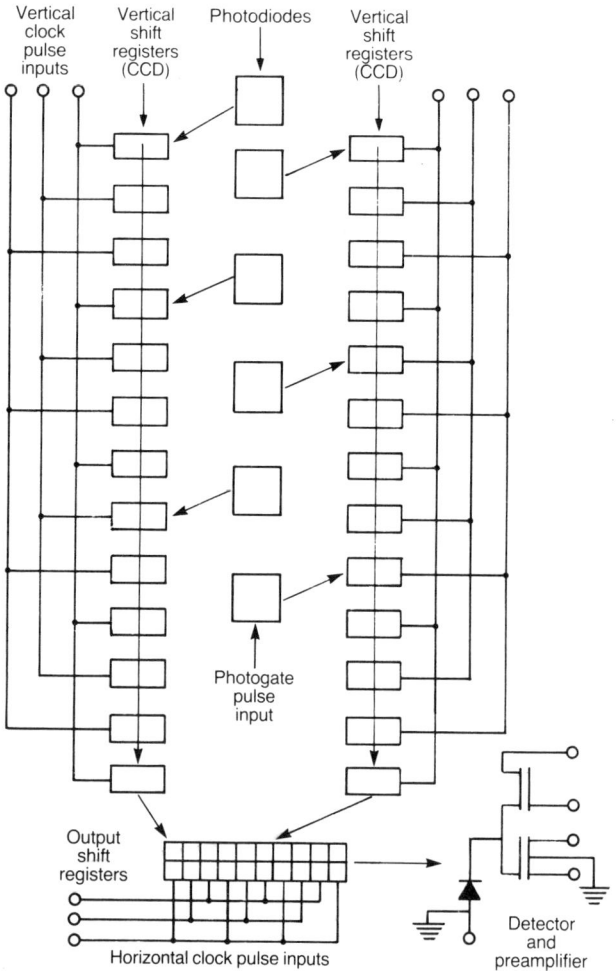

Figure 6.10. Schematic representation of a CCD image sensor (Spectrum, January 1975, p. 55)

Figure 6.10 illustrates a CCD image sensor consisting of a small panel measuring approximately 30×12 mm with 10 000 photodiodes together with 100 columns of CCD shift registers. When an image is focused on the photodiodes they respond by generating electron–hole pairs. A photogate clock pulse is then applied as a bias potential between the diodes and the vertical shift registers which permits the electrons accumulated by the action of the light to transfer to the shift registers, which also receive a pulse having a different voltage value on every third scan. To ensure continuity, only half the diodes discharge into the corresponding vertical shift registers while the other diodes continue to charge through the action of light. In this manner alternate batches of electrons are injected to the left and right registers. The batches of electrons are then transferred vertically through the shift registers to two-element horizontal output registers which interleave alternate batches from the left and right vertical registers to restore the proper sequence of image signals. These output signals are charge signals and hence have to be converted into voltage signals before they can be amplified. For conversion a reverse-biased diode is used. The potential developed across this component changes linearly when an electron packet is delivered to it and this change in potential is then used to bias the gate of a MOSFET preamplifier.

7 Fabrication of semiconductors and integrated circuits

The evolution of electronic technology achieved over the past two decades has profoundly increased the capabilities of electronic devices. Smaller and smaller electronic components perform increasingly complex functions at high speed and low cost. The basis of this advancement can be said to date from 1960 with the development of a process for the fabrication of silicon transistors. Known as the planar process it is carried out on wafers of silicon and involves a combination of oxidation, selective oxide removal and heating to introduce dopant materials by a diffusion process. Although it is now the principal method of fabrication for semiconductor devices and integrated circuits, virtually every stage of fabrication is either in the midst of significant advances or on the verge of one. The preliminary step involves the preparation of a silicon wafer on which the devices are fabricated. As the properties of semiconductors are extremely sensitive to the presence of impurities the preparation of the wafer requires that it must be refined to a very high degree of purity. In fact the impurity level must not exceed one part in 10^{10}. The basic steps involved in wafer preparation are (1) extraction of the material from ore, (2) purification, (3) crystal growth and (4) wafer processing.

Metal extraction

Germanium

Although as abundant in the earth's crust as such common metals as zinc or lead there are only a few minerals in which germanium is an important constituent and these occur only in small deposits. It is economically recovered from zinc ores, with which it is frequently associated in small amounts, as a byproduct in the recovery of the primary metal. The element is separated from contaminating materials by distillation of the tetrachloride from strong hydrochloric acid solutions. Hydrolysis of the tetrachloride with water results in the precipitation of germanium dioxide.

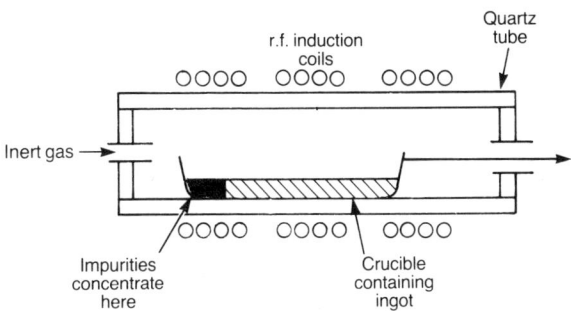

Figure 7.1. Zone refining. An ingot of the material contained in a crucible is pulled through an induction-heated quartz tube. A molten zone containing the impurities travels through the solid ingot to the end of the ingot that solidifies last

Reduction of dioxide to metal is carried out by heating in a current of hydrogen at a temperature of 600–650 °C. To obtain the pure metal necessary for semiconductor applications the material is usually subjected to a process known as zone refining, which is based on the segregation of impurities between a molten and a solid part of the material under the influence of heat (Figure 7.1). By selectively heating an ingot of the metal it is possible to produce a molten section or zone which can be moved from one end of the ingot to the other end. Impurities tend to concentrate in this molten section as it passes through the ingot and hence impurities can be swept to one end of the ingot, leaving the rest of the ingot in a pure condition. By breaking off the impure end of the ingot and repeating the process, purification to a high degree may be effected.

Silicon

Silicon is of course extremely common for it is the most abundant solid element in the earth's crust being present in the form of oxide and silicates. It is produced in large quantities by the reduction of silicon with coke in an electric furnace according to the overall reaction $SiO_2 + 2C = Si + 2CO$. The 98-99% product is then subjected to the action of chlorine, the silicon tetrachloride so formed being separated from impurities by fractional distillation. Reduction with zinc vapour at a temperature of 1000 °C results in the deposition of pure silicon. Zone refining as used for germanium is not effective in purifying silicon. The high temperature required to melt silicon, 1420 °C as compared with the melting point of germanium of 958 °C, causes problems chiefly because silicon in the molten state is easily contaminated by impurities in the crucible containing the metal.

Crystal growth

Semiconductor materials prepared by the above methods result in a polycrystalline of irregular crystalline structure. Such a material

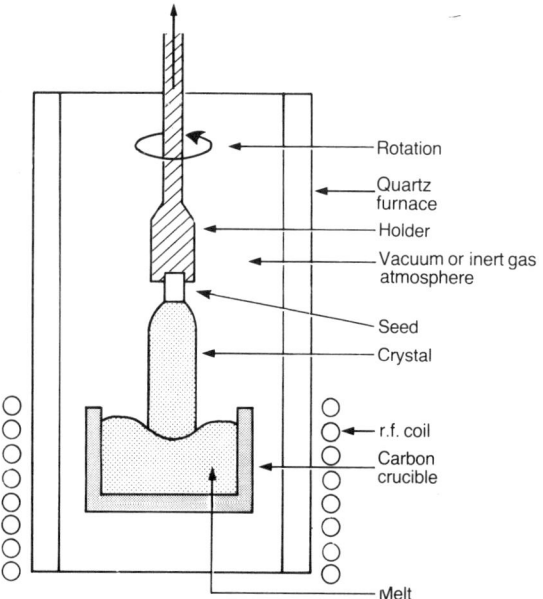

Figure 7.2. Production of a single crystal by the Czochralski crystal-pulling technique

tends to reduce the mobility of carriers. For it to be useful as a semiconductor material it must be grown as a single crystal. The process known as crystal pulling is usually combined with the addition of controlled amounts of acceptor or donor impurities to produce the required electrical characteristics.

A widely used method for single crystal growth is the Czochralski technique (Figure 7.2) in which a crystal is grown by the slow withdrawal of a sedd crystal from the molten liquid contained in a crucible and kept just above the melting point. The seed is withdrawn at a rate, typically 1 to 2 mm per minute, sufficient to keep the liquid-solid interface near the melt surface. The slower the

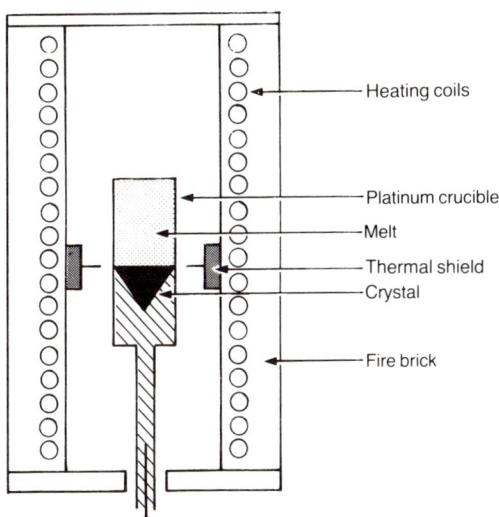

Figure 7.3. Bridgman–Stockbarger crystallisation technique. Molten material is lowered from a hot to a cold zone

rate the more perfect the crystal. Since the seed is at a lower temperature than the melt, heat flows from the melt to the seed, the atoms of the melt solidify on the seed and arrange themselves so that they may have the same crystal structure as the seed crystal. By addition of carefully controlled amounts of impurities to the melt the resulting crystal may be either *n*- or *p*-type. The dopants commonly used are phosphorus or antimony for *n*-type and boron for p-type. Single crystals 7–10 cm in diameter and 30–50 cm long can be pulled from the melt. For industrial purposes Czochralski crystal-pulling devices are usually automated on a large scale. Another crystal growing method uses the Bridgman–Stockbarger technique illustrated in Figure 7.3. Here the substance is enclosed in a sealed crucible and slowly lowered into and out of the hot zone of a furnace maintained just above the melting point of the material. A seed forms at the tip which is coolest and leads to the formation of a nucleus which grows as the crucible is lowered. The rod-like crystal is then sliced by a diamond-tipped rotating saw to yield wafers approximately 0.50 mm thick which are then polished and chemically treated to remove any surface contaminations (Figure 7.6).

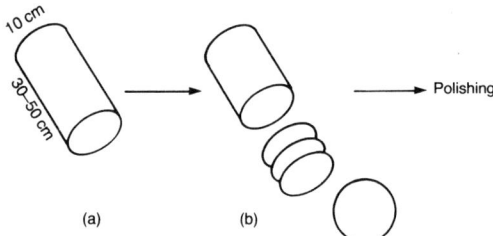

Figure 7.6. Simplified outline of the production of silicon wafers which serve as substrate for transistors and ICs. (a) Silicon ingot prepared by the Czochralski crystal-pulling technique is sliced into wafers 0.5 mm thick (b) which are then smoothed and polished. The wafers are then processed as outlined in Figure 7.5.

Production

Two basic processes are in operation for the production of semiconductors, namely the alloy process and the planar (diffusion) method. The first was in extensive use during the 1950s and is still used in the fabrication of germanium transistors, although for silicon devices it has now been superseded by diffusion methods.

Alloying

The method of construction for a p–n junction type semiconductor is illustrated in Figure 7.4. A pellet of p-type material, usually indium, is heated in contact with *n*-type germanium to a temperature of about 500 °C. The indium melts and dissolves some of the germanium from the wafer. After cooling and recrystallisation, regions of *n*-type material form as shown in the diagram. In the case

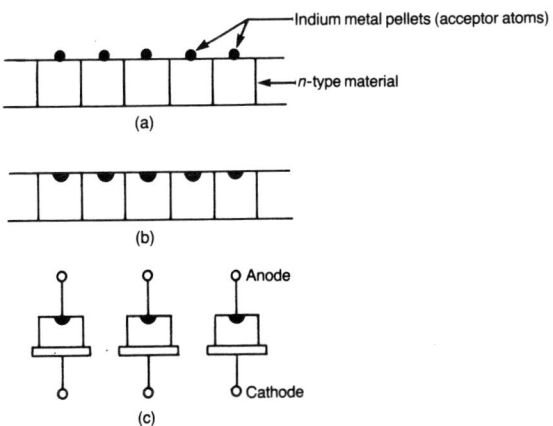

Figure 7.4. Fabrication of a p–n junction by the alloy process. (a) Pellets of indium are heated in contact with n-type germanium; (b) indium diffuses into the germanium; (c) the material is then sectioned, bonded to a metal base, leads are fitted and the whole is encapsulated

of a transistor, two pellets of indium would be used one on each side of a germanium wafer kept in place by a jig and the same procedure followed. This operation would form a *pnp* transistor.

Planar process

This undoubtedly is one of the most important processes in modern semiconductor technology for it has revolutionised design, manufacture and application. The planar technique invented in 1960 was first used for the fabrication of silicon transistors. Later it was discovered that the process could be expanded to include resistors, capacitors and diodes and hence complete functional circuits could be incorporated within a single substrate. The process is now universal in the manufacture of silicon integrated circuits. The name planar is derived from the fact that the fabrication is carried out in a wafer of silicon through a series of operations that require access to only one surface (plane) of the wafer.

Planar transistors

In contrast to the alloying process which involves a liquid phase the planar method uses diffusion, a gaseous phase being used for the introduction of the dopants that create p- and n-regions. The process involves a number of successive operations, the sequential steps being outlined in Figure 7.5. These steps are: (a) masking, (b) photolithography, (c) diffusion of dopants, (d) metallisation and (e) sectioning into individual dice or chips.

Masking. Wafers of n-type silicon are exposed to steam in a furnace at a temperature of 1200 °C. An oxide film about 0.4 to 0.6 μm thick is formed in an hour. The dominant role of silicon as the material for semiconductors is attributable in large part to the property of this oxide layer. It is an excellent insulator and is hard and durable and is particularly important because it can act as a mask for the selective introduction of dopants.

Photolithography. This is a photoengraving process for the formation of patterns on the surface of the oxidised wafer. The oxidised wafer is coated with a photosensitive material known as a photoresist. This is a light sensitive lacquer composed of a liquid solution of resin in an organic solvent which hardens when exposed to light. A mask containing the desired pattern is then placed over the wafer and illuminated by an intense light source which fixes certain areas. The unfixed photoresist is then dissolved by a special solvent. The next step involves a hydrofluoric acid etch to remove the silicon dioxide from the unfixed areas, the rest of the oxide being protected by the insoluble photoresist. The insoluble photoresist is now removed by another chemical treatment. The oxide layer has now open areas called windows through which selective impurities can be introduced into the underlying silicon.

Dopant diffusion. There are two techniques for selectively introducing dopants into the silicon crystal wafer; diffusion and ion implantation. In the diffusion process the wafer is placed in a furnace which contains a source of the impurity atoms. If a *npn* device (the most common in ICs) is required the p-type dopant will be boron tribromide or phosphorus oxychloride creating a *p–n* junction. A second sequence of photolithographic oxide etching and

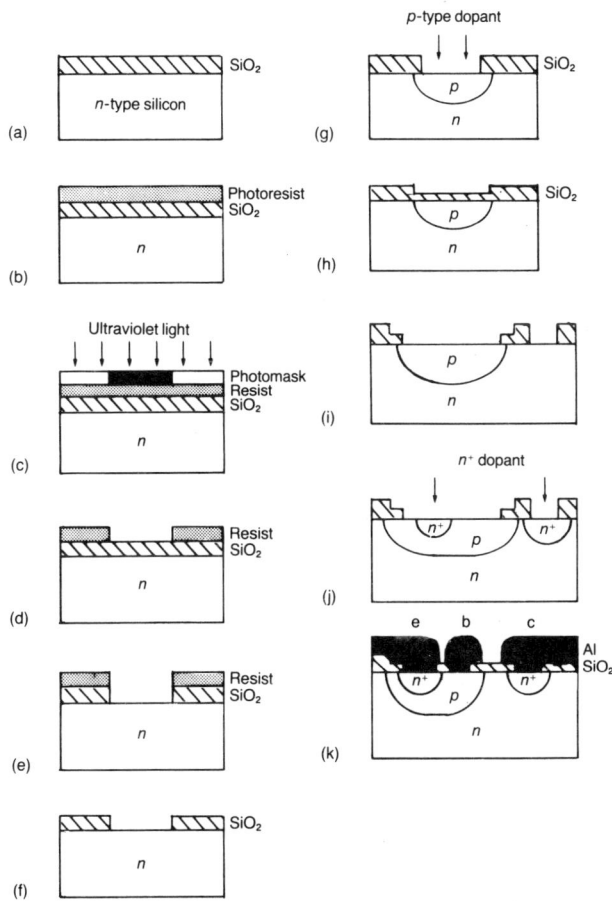

Figure 7.5. Planar process. (a) Oxidation; (b) application of photoresist; (c) exposure to UV light through the photomask; (d) application of a special solvent results in the removal of the unexposed photoresist; (e) hydrofluoric acid dissolves the oxide layer wherever it is unprotected; (f) the photoresist is removed by another chemical treatment leaving the photoresist pattern in the silicon substrate unaffected. (g–k) Repetition of the photolithographic method for production of an npn transistor

diffusion steps follow with an n^+-type dopant. The whole surface is oxidised again and windows are cut by the photoresist technique to give access to the electrodes. A thin layer of metal (aluminium) is then deposited over the whole top surface and those areas required for electrode connections are etched away. The wafer is now diced, base and emitter leads are welded on, the dice are bonded to a metal header which forms the collector connection (Figure 7.7) and finally the device is encapsulated in plastics to prevent the ingress of dirt and moisture and protect against mechanical abrasions. With the planar technique production tolerances can be controlled to within a micrometre.

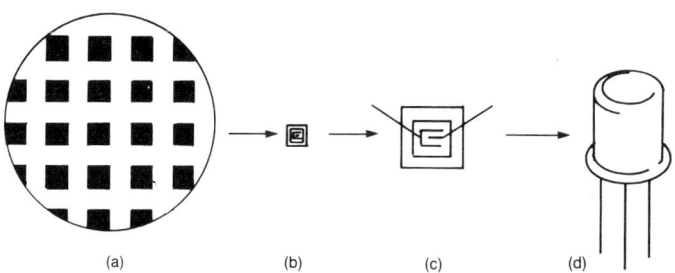

(a)   (b)   (c)   (d)

Figure 7.7. Following the photolithographic process the wafers are (a) scribed (sectioned) with a diamond-tipped tool into individual chips (b). Base and emitter leads are attached (c), the chips are bonded to a metal header and encapsulated (d)

It should be mentioned that work to develop a similar process using germanium and incorporating germanium oxide as the surface mask has proved unsuccessful because of the unsuitable nature of the oxide. It has, however, been found practicable to deposit a layer of silicon oxide on the surface of the germanium, and then proceed with the planar technique as before.

Ion Implantation. The other selective doping process, ion implantation has been developed as a means of introducing dopants at a lower temperature. In this method the doping material is ionised and accelerated in an electrostatic field and fired at the material to be treated, the areas into which the ions are fired being determined by masking the areas with aluminium foil (Figure 7.8). The ions penetrate the target and lose their energy in collision with the atoms and electrons in the solid until the rest energy (thermal energy) is reached. The advantages of this method are that the quantity of impurities introduced, the depth of penetration and therefore the concentration in the crystal are parameters which can be controlled precisely and independently through the intensity and energy of the ion beam. In thermal diffusion it is possible to act on the diffusion time and temperature, but not independently. Ion implantation takes place at a much lower temperature (500–700 °C as against over 1000 °C), the ions penetrate the crystal as a directional beam and in

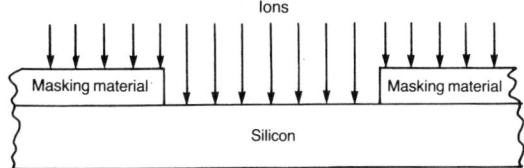

Figure 7.8. Ion implantation

this way the doped region is defined with remarkable precision contrary to what happens in diffusion where the isotropy of the process causes leakage at the edges of the mask which delimits the doping zone. The main disadvantage is that the energised ion penetrating the crystal lattice collides with the atoms of the solid, breaking the bonds so that the regularity of structure is lost together with its characteristics and properties. In extreme cases the solid changes from a crystalline to an amorphous structure. It is possible, however, to correct most of the irregularity by annealing at a moderate temperature.

Gallium arsenide

In recent years this semiconductor material has had a major impact on a wide range of devices. These include oscillators (Gunn and avalanche effect), amplifiers (Gunn and FETs) and Schottky-barrier diodes (for use as detectors, mixers and varactors). In addition it has found use in light-emitting diodes (LEDs) which make up the numerical displays of some electronic calculators. Its electron mobility and band gap are high compared with those of silicon and germanium and by the introduction of suitable dopants a wide range of electrical conductivity values can be attained. For instance it has about six times greater conductivity than silicon doped to the same level with n-type impurities. Further it can be produced as a semi-insulating material with resistivity of 10^6 Ω cm^{-1}. Thus it can be used as a device substrate on which contacts are replaced to minimise parasitic capacitances. The material can also be used as a dielectric-propagating medium in amplifiers making use of travelling wave interactions or in monolithic gallium arsenide microwave circuits. These uses together with the discovery of the Gunn effect that occurs in gallium arsenide because of its particular band structure have stimulated interest in this material. Like other semiconductor devices gallium arsenide components are constructed with their active regions fabricated within a layer grown epitaxially on a monocrystalline substrate of the same material. The required properties in the epitaxial layer are obtained by doping with an appropriate impurity. Epitaxial thicknesses may range from less than 1 μm to many tens of μm depending on the particular device. The major emphasis in the study of epitaxial growth of gallium arsenide has been in the preparation of material suitable for Gunn-oscillation devices. Epitaxial layers for this purpose must be of high purity combined with a high electron mobility.

Preparation

Two general methods are available. The first is chemical vapour deposition, the second solution growth from a gallium melt. In the former, high-purity hydrogen is passed together with arsenic trichloride into a silicon reaction tube in a two-zone furnace containing gallium heated to 800–850 °C. Chemical reaction takes place to form gaseous gallium and arsenic compounds, the gaseous reaction products pass into a second zone maintained at a lower temperature (750 °C) where deposition of gallium arsenide occurs on a gallium arsenide substrate. The overall reaction can be represented as follows:

$$2\,\text{GaCl} + \tfrac{1}{2}\text{As}_4 = 2\,\text{GaAs} + 2\text{HCl}$$

Dopants such as sulphur, tellurium (n-type) cadmium (p-type) are added to the gas stream.

In the other technique, liquid phase epitaxy takes place on a gallium arsenide substrate slice which is lowered into a crucible containing a saturated solution of gallium arsenide in molten gallium at around 700 °C. This is then cooled and gallium arsenide is deposited on the substrate surface. Dopants such as tin are added directly for the growth of *n*-type material. At present, both vapour and liquid phase deposition processes are used. It would seem that the liquid-phase method may be preferred for the growth of thick, lightly doped layers but the development of the vapour system has resulted in greater versatility, better doping control and automation of the various process operations.

Field effect transistors

The fabrication sequence for a FET follows the same pattern as that for bipolar transistors (Figure 7.9). In a typical *p*-channel enhancement type MOST a substrate of *n*-type silicon is used with *p*-type source and drain regions diffused into the substrate by the use of masking devices. Over the channel between the source and drain a thin layer of silicon oxide is formed. For the electrodes windows are opened directly over the *p*-regions by an etching process.

The oxidation step is fundamental for it serves not only to define accurately the final gate oxidation thickness, but also to determine the surface-state density of the oxide–silicon interface. This latter is crucial to satisfactory device performance since its regulation is a dominant factor in the value and variation of the minimum value of the gate voltage just sufficient to cause channel formation and constitutes the gate threshold voltage.

For a MOST device the critical dimensions are the thickness of the oxide layer under the gate electrode and the distance separating the source and drain regions. The reason being that the sensitivity of the device's response to a gate voltage varies inversely as the thickness of the oxide layer. Typically the thickness of the oxide layer is 0.1 μm and the spacing between the source and drain is 7.5 μm.

Integrated circuits

Integrated circuits is a technology of highly miniaturised electronic circuitry, and its practical application can be said to have been the most significant technological development of the 1960s. Their first

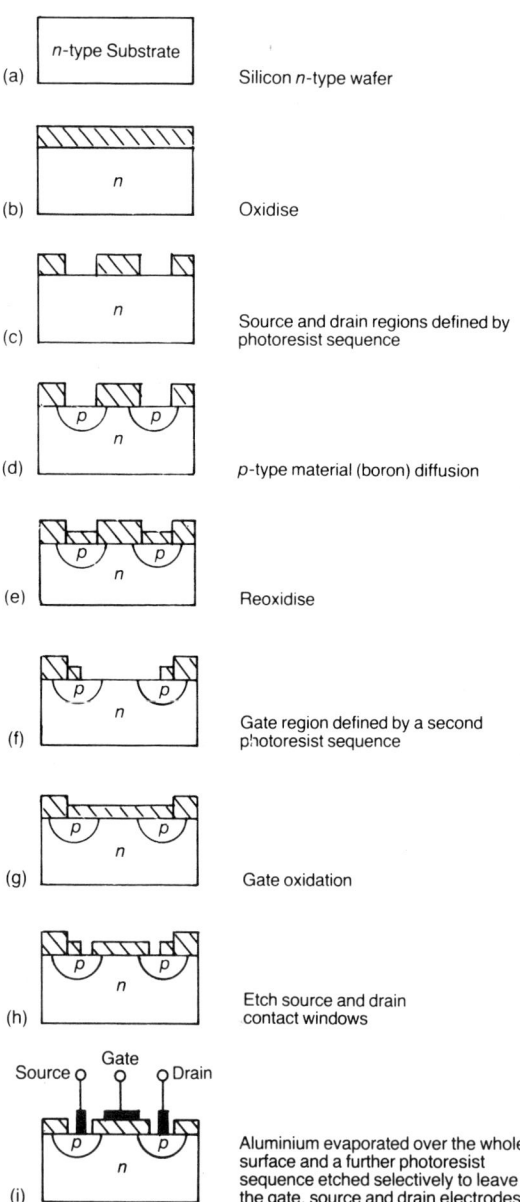

Figure 7.9. The fabrication of MOST (enhancement type)

major use was in the field of computers but it was not long before this new technology found its way into many other applications. IC technology introduced new methods into the fabrication and use of semiconductors and eliminated the need for separate (discrete) electronic devices such as resistors, capacitors and transistors as the building blocks of the electronic circuit. In their place were substituted tiny chips of silicon no larger than a pin head whose functions were not those of a single component but which contained dozens of transistors, resistors, capacitors and other electronic components all interconnected to perform the task of a complex circuit. In other words, individual component parts become less identifiable and were integrated into an overall circuit function. This made it possible to build complex electronic assemblies in a batch production process, in which a number of circuits were processed simultaneously during the course of manufacture. The advantages of ICs over their discrete counterparts are: greatly reduced size and weight; small power consumtpion; and improved reliability and performance at lower cost. In short, it became possible to build complex electronic assemblies in a batch production process that effectively increased equipment reliability while reducing equipment cost.

Classification

Integrated circuits may be classified according to method of fabrication (Figure 7.10). The basic semiconductor IC is a silicon monolithic structure (monolithic meaning formed from a single block) in which all components both passive and active are formed

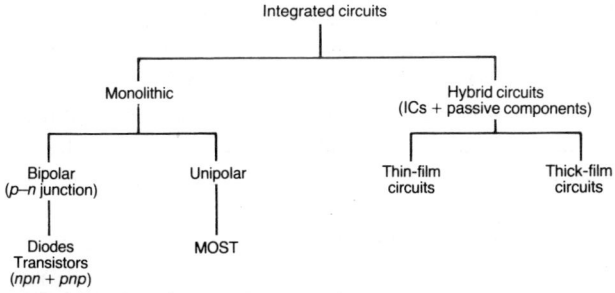

Figure 7.10. Classification of integrated circuits

simultaneously in a single crystal wafer of silicon by forming zones of p-type and n-type material in the crystal in a particular sequence by the diffused planar technique. A second type of IC is the hybrid which in contrast to the monolithic IC is formed by a film process on a suitable substrate and is limited to the fabrication of passive components. They are not fabricated in a single crystal of silicon but rather from whatever materials are best suited to the individual component function. Two film technologies are used: (1) thin films which use vacuum deposition of various materials to form passive components and (2) thick films which use specially prepared inks to form passive components. The required active devices are added separately to the film in the form of discrete device chips.

Monolithic ICs

In monolithic integrated circuits fabrication commences with a silicon crystal 30–50 cm long and 10 cm in diameter which is sliced into hundreds of wafers about 0.5 mm thick. These are then converted into *p*-type material and steam passed over them at 1200 °C to form a thin surface layer of silicon dioxide on the exposed surface. By photolithographical methods 'windows' through which the doping materials can penetrate are etched into the oxide layer.

The next operation involves the formation by diffusion of an *n*-type region for each component as shown in Figure 7.11. This is followed by diffusion of a *p*-type impurity (as for instance boron) for the transistor base, the diode anode, resistor (which is simply a layer of *p*-type silicon) and the capacitor. Diffusion of a high concentration *n*-type (n^+) impurity such as phosphorous forms the transistor n^+-type emitter region and at the same time the collector n^+-region, the diode cathode and the capacitor n^+ electrode.

Resistors are formed by the diffusion of a thin film of *p*-type silicon into an isolated island of the *n*-type epitaxial layer — Figure 7.12a. The diffusion is usually accomplished simultaneously with the base diffusion of transistors. They are usually designed as a narrow strip with ohmic contacts at the two ends, as illustrated in Figure 7.12b, which have been etched through the overlaying silicon dioxide. Current flowing through the device is confined to the film by maintaining it at a negative potential with respect to the *n*-type island so that only reverse current flows across the *p–n* junction.

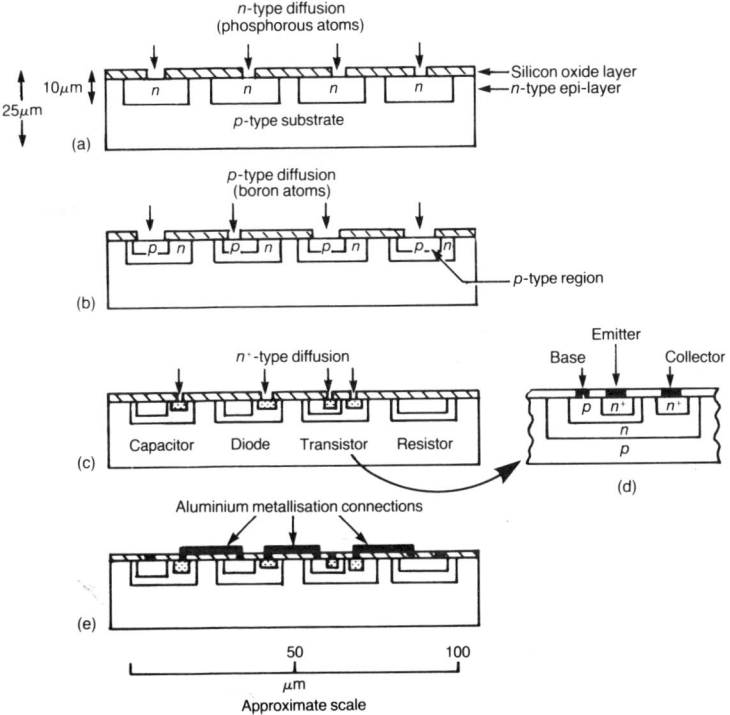

Figure 7.11. Sequence of diffusion steps for the manufacture of ICs. (a) n-type diffusion; (b) p-type diffusion; (c) n^+-type diffusion; (d) cross section of IC transistor; (e) metallisation

Capacitors are formed in a similar way to resistors during the n^+-type transistor emitter diffusion which forms one plate. A layer of aluminium deposited on top of the SiO_2 (which acts as the dielectric) forms the other plate (Figure 7.13). As the total area occupied by the device is very small, the values of capacitance available in ICs is seldom more than about 5 pF.

Inductors, as already mentioned, cannot be made by the diffusion technique. Hence if an induction impedance is required it must be supplied as a discrete component.

The final operation consists of depositing the interconnecting metallisation necessary to connect the components into the required circuit arrangements. Aluminium is commonly used since it adheres well to both silicon and silicon dioxide and is evaporated as a thin film on the SiO₂ layer. A cross section of part of an IC showing both active and passive components is shown in Figure 7.14.

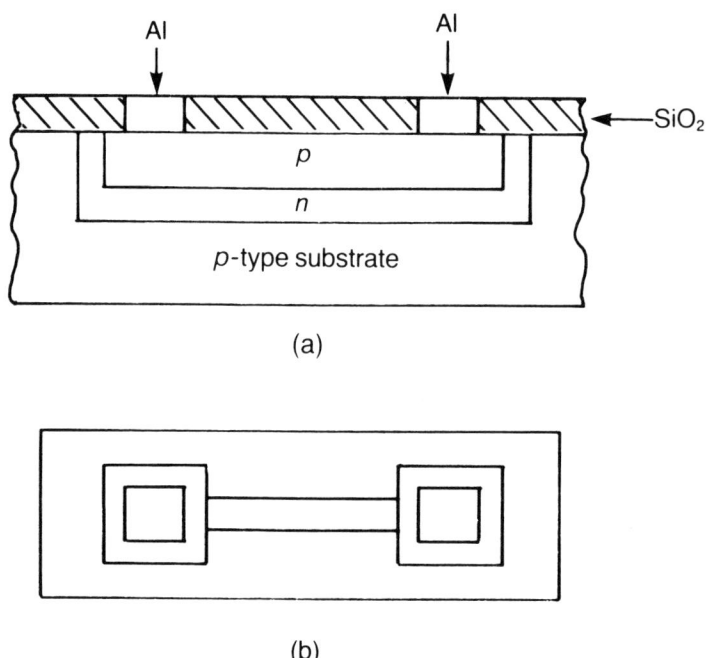

Figure 7.12. Cross section (a) and plan (b) of a diffused-type IC resistor

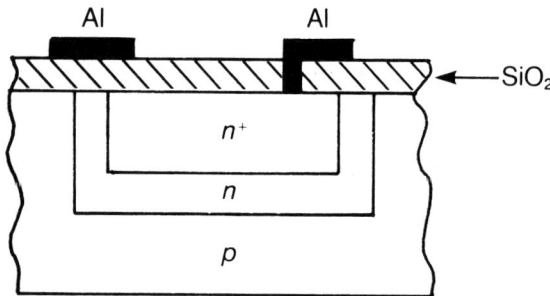

Figure 7.13. Cross section of an IC capacitor

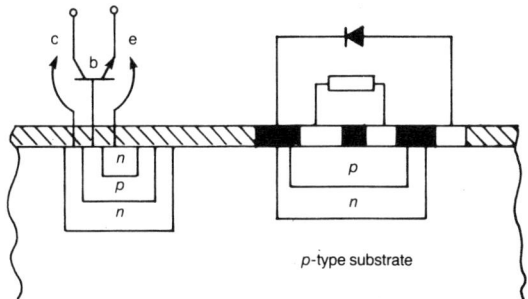

Figure 7.14. Cross section of an IC incorporating a transistor, a resistor and a diode formed by the mask-etch-diffusion planar technique

Photolithography

The fabrication of integrated circuits requires a method for accurately forming patterns on the silicon wafer. Photolithography is used for this purpose. The process has already been outlined (p.124) but requires amplification as applied to integrated and microelectronic circuits. It can be defined as a process by which a microscopic pattern is transferred from a photomask to the silicon wafer. The fundamental material is the photomask which serve as the stencil for delineating the circuit components and interconnections. In the preparation of the masks a large-scale master drawing which may be 500 times larger than the actual circuit is first made with the transistors, diodes, resistors and capacitors accurately dimensioned and positioned. This is then reduced photographically to the size of the wafer. Thus, the final size of the mask may involve a reduction to one-five-hundredth of the original size. In the case of ICs identical circuits are reproduced many hundreds of times on the mask by a camera using a process known as 'step and repeat' to yield a set of final-size master masks from which a large number of working plates are copied. A working copy is made which may be either a fixed image in an ordinary photographic emulsion or a more durable pattern etched in a chromium film on a glass substrate. Perfect alignment of the photomask with the silicon wafer is secured by means of a machine which accurately positions the mask in contact with the wafer.

Once processing is complete the wafer will contain many identical circuits. Separation into individual circuit chips is achieved by scribing with a diamond tool and then mechanically breaking the wafer along the scribed lines into rectangular dice. The individual die is then mounted on a metal base known as a header and the whole encapsulated in plastics to produce the familiar 'chip', electrical connection being made possible by external leadouts attached to the header for connection to the power supply and any external components. By far the most widely used package is the dual-in-line type (Figure 7.15c).

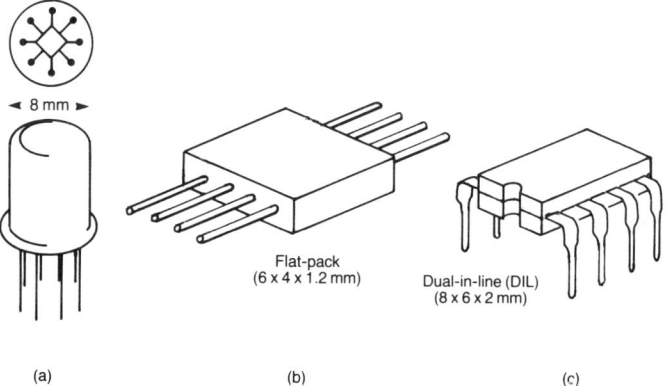

Figure 7.15. IC assemblies. (a) Pin type TO-5 package; (b) eight lead flat pack; (c) dual-in-line (DIL) package, the most widely used assembly

Special isolating techniques are used to prevent random leakage currents between elements from affecting the performance of the circuit. The size of a typical monolithic logic element for a computer may be as small as $1 \times 1 \times 0.3$ mm and the only connections that need be made are those to the power supply and input and output units.

During the complex treatment there are many points at which chips may be damaged or destroyed, this wastage reducing the yield by as much as 80%. However, because 500 or so complete circuits can be produced simultaneously on a single wafer, the total output is so large that the final cost per circuit is very low. Identification of the damaged chips is by the use of electrical and microscopic inspection techniques. Any consequent rejection is thus made possible before the expensive assembly operation.

Although the fabrication of ICs by the planar process appears involved and complex, it must be remembered that the manufacture entails the simultaneous production of hundreds of circuits side by side on a single wafer. Mass production is taken a stage further by processing as many as 100 wafers together in a batch. Hence the cost of labour and equipment is shared by thousands of circuits, making possible an extremely low unit cost. Less than 10% of the cost of electronic equipment is in the ICs themselves. This explains the reason why today most of the electronic industry depends in some way or other on ICs.

Future developments

The technology of lithography which has been stable for many years is now undergoing change. The conventional photolithographic process where the mask is put in contact with the silicon wafer is being replaced by an optical projection method in which the patterned mask is projected on to the wafer. This development is capable of resolving finer structures giving a reduction in size leading to an increase in the density of ICs on the wafer. The smallest features that can be formed by present lithographic processes are ultimately limited by the wavelength of light. Present technology can routinely reproduce elements a few μm across. To form sharp images of still smaller structures new lithographic and fabrication techniques will be required. Circuit patterns will have to be formed with radiation having wavelengths shorter than those of light. With this in mind a number of advanced techniques are now under development. Electron beams and X-rays with wavelengths in the nanometre (10^{-9} m) range are capable of producing extremely fine features. In electron-beam lithography a fine beam of electrons scans the wafers to expose an electron-sensitive resist in the desired areas. The beams have a high electron energy (5–30 keV) so that diffraction effects are negligible. This allows better definition of details in the range 0.5–2.0 μm than is possible with photolithographic procedures. It is possible to focus beams to very small diameters (3 nm) and this allows patterns to be made with very fine detail. The ability to make such fine detail leads to increase in the packing density of the elements of ICs with consequent reduction in their size and cost. X-ray lithography is a form of contact photolithography in which a modified form of X-rays is substituted for ultraviolet radiation. It promises to be a sub-micrometre lithographic technique that is potentially high in throughput and relatively inexpensive.

Epitaxy

Although the details discussed above include the basic operations of the planar process there is another technique for the deposition of silicon that is important in the manufacture of ICs. This is the epitaxial process, the term being derived from the Greek word 'epi' meaning upon and 'taxos' meaning arranged.

In the planar process, the impurities are diffused into the substrate but epitaxial growth involves deposition of a layer on to the substrate. An epitaxial layer is a single crystal silicon layer (Figure 7.15), deposited from a gas stream, in which transistors and

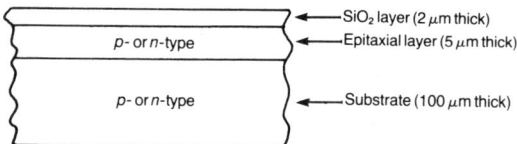

Figure 7.16. Epitaxial deposition. The chip is produced by growing a thin high-purity layer (high resistivity) of semiconductor material on a heavily doped substrate (low resistivity). By the addition of suitable impurities p- and n-type regions can be formed in any sequence and with any desired concentration gradient within the layer. In this way certain characteristics, for example, electrical conductivity can be varied more or less at will. Epitaxial deposition is a flexible process with wide applications in the manufacture of ICs

other circuit components will ultimately be fabricated. This layer may be of higher or lower impurity concentration than the substrate or of the same or opposite conductivity. Thus an n-type layer may be grown on a p-type substrate or a lightly doped n-type layer on a n^+-doped material. Further, by the introduction of suitable impurities during the doping process p- and n-type layers can be formed in any sequence and with any desired concentration. Thickness can also be controlled over a wide range and in fact as transistors become smaller it is also necessary to make the epitaxial layer thinner and these have decreased steadily from 10–15 μm thick to in some extreme cases 5 μm.

In addition to being extremely flexible, epitaxy eliminates the long diffusion required in the planar process and this shortened process time has undoubtedly caused the yield of ICs to be increased.

Production
Epitaxial silicon films are grown by vapour phase deposition on a wafer substrate heated to 1100 °C in a cylindrical quartz tube encircled by a r.f. induction coil heater. The wafers are placed in graphite boats and inserted in the furnace. The gases required for the reaction are introduced via a control valve. For a silicon epitaxial film a gas mixture composed of silicon tetrachloride and hydrogen is introduced, the tetrachloride dissociates and silicon atoms are deposited in accordance with the equation $SiCl_4 + 2H_2 = Si + 4HCl$. By addition of suitable dopants to the gas mixture, n- or p-type semiconductors can be obtained or by the addition of small amounts of oxygen or ammonia into the gas stream, silicon oxide or silicon nitride insulating layers can be formed.

Thin and thick film circuits

The silicon IC which appears to approach the ultimate in miniaturisation has together with its many advantages a number of limitations. Specialised circuits requiring high power, high operating frequency and low component tolerances cannot be produced as silicon ICs. Resistance and capacitance values are strictly limited and they cannot be adjusted individually. Passive elements in a silicon IC must be isolated by reverse biased junctions which lead to undesirable stray capacitance. Since all the active and passive devices are fabricated in the semiconductor wafer, parasitic losses and spurious coupling between the various elements on the same substrate can seriously limit the circuit performance.

Hybrid circuit

To overcome these limitations one obvious method is to produce active and passive components on separate substrates. In other words the active devices (transistors and diodes) are produced in the form of discrete chips and then connected to the passive components. This gives rise to what is known as the hybrid (multi-chip) circuit which uses the advantages of the monolithic silicon IC yet allows the design freedom of the conventional circuit. To take advantage of the size reduction offered by ICs, passive components (resistors and capacitors) have to be reduced in size. Miniaturisation of passive components is effected by a process known as film technology in which passive components are deposited in the form of a film on to a flat substrate. Interconnections are thick metal films deposited in the required pattern which conncct these elements to each other and to the active devices needed to complete a functional circuit. Thin film resistors are metal strips usually about 0.5 to 2.5 μm thick, resistances up to 500 kΩ can be made on one square centimetre of substrate. Capacitors are sandwiches of two metal film electrodes separated by a dielectric film, values of up to about 0.1 μF cm^2 being obtained. It may be said that the hybrid microelectronic technology supplements the silicon IC in the miniaturisation field. The approach is particularly useful where a relatively small number of identical circuits are required since the design layout and fabrication cost of hybrid circuits is less than for a similar silicon IC.

Film technology is of two basic types — thin and thick film, the former approximating 0.025 to 2.5 μm in thickness, the latter 25–50 μm.

Thin film circuits

Thin film passive circuits are fabricated by the deposition of a metal film on to a glazed alumina substrate. Materials for resistors include tantalum, Nichrome and tin oxide, for conductors copper and gold, and silicon monoxide for insulation layers. The required resistor–conductor pattern is obtained by the use of photolithographic masks and etching techniques. Fabrication of thin films is either by vacuum evaporation or sputtering.

Vacuum evaporation
Material contained in a suitable crucible is vaporised in a vacuum chamber and deposited on to a substrate a few centimetres distant. The substrate is located over a mask composed of a heat resistant metal, molybdenum, which defines the pattern to be evaporated. Control of the substrate temperature, degree of vacuum, rate of

evaporation of the material and the thickness of the deposit are automatically regulated. Total cycle time for the deposition of a film 1 μm thick is about 12 minutes.

Sputtering

When the metal to be deposited is particularly refractory it is convenient to use an alternative technique called sputtering. Instead of being heated to the point when it vaporises the metal is subjected to intense bombardment by the ions of a gas. Argon is generally used as the ions are inert and have a relatively high mass. Atoms ejected from the surface of the target metal are sputtered as a thin film on to the substrate. The target metal is made the cathode and a potential of several thousand volts is applied between it and the substrate which are separated by a few centimetres. Sputtered films are more strongly adherent than evaporated films because the arrival energies of sputtered atoms are much higher. A disadvantage of using sputtering is that the rate at which the film can be deposited is slow. Typically films grow in thickness at rates around 10^{-8} cm s^{-1}. By comparison rates of about 10^{-6} cm s^{-1} are common with evaporation.

Thick film circuits

Thick film technology is based on a process known as screen printing used in the printing industry in which ink is squeezed through a mesh stencil on to the surface to be printed. In the case of thick film technology these inks are basically dispersions of glass particles and precious metal salts suspended in an organic vehicle. The mask which is placed in intimate contact with the substrate is composed of a woven mesh of stainless steel or nylon the pattern being defined by slots and holes. The ink is forced through the apertures leaving them filled with ink. Any excess is scraped off the surface, the mask is then removed, the ink dried and then fired at a temperature of about 800 °C. On firing the organic vehicle decomposes leaving the material dispersed on the substrate, the glass particles bonding the film to the substrate.

The circuit may then be protected with a glaze which may be printed to leave windows where connections to the conductors will be required. Resistor material is commonly a mixture of palladium, palladium oxide and a borosilicate glass frit dispersed in the organic vehicle. A wide variety of resistor values and patterns can be obtained with little process variation. For example by varying the ratio of metal to glass in the inks, resistance values ranging from 1 Ω to 500 MΩ can be achieved.

Fabrication of capacitors is achieved by a similar technique, an insulating layer being interspersed between two layers of conductors. Little success has been achieved in the fabrication of inductors, the helical coils and ferromagnetic core materials have so far proved incompatible with film technology.

8 Analogue circuitry

Signals are the physical representation of a message and are therefore carriers of the information to be processed. The construction and mode of operation of control systems are therefore decisively influenced by the type of signals used. The manner in

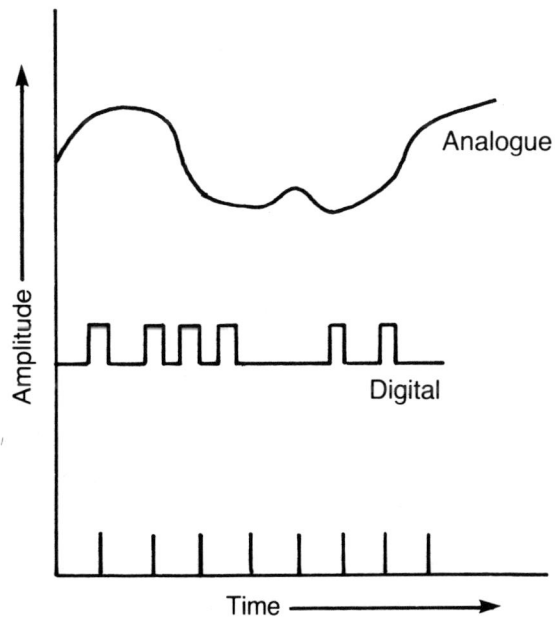

Figure 8.1. The two basic types of signal

which a signal may carry information falls into two broad classes, analogue and digital (Figure 8.1).

The word analogue is derived from the word analogous, meaning 'similar to'. Since all electric circuits use voltages and currents to carry out their function, analogue circuits use currents and voltages that are analogous or similar to whatever quantity is being described, measured, generated and/or controlled.

A feature of an analogue signal is that the information-carrying characteristics of the signal (for example, amplitude, phase) are continuously variable between certain limits. The fundamental feature of analogue signals is the condition that they can assume any desired amount within the range of their limits and they may be required to perform almost any combination of operations or signals. These operations may include addition, subtraction, multiplication etc.

Digital signals on the other hand are quantified both in amplitude and time and signals are said to be digital if only a restricted number of values of the information carrying characteristic are defined. Digital is derived from the English word digit and characterises the numerical representation of analogue quantities by binary signals in which there are only two digits 0 and 1, that is they are restricted to two levels.

Analogue signals are often referred to as linear and used to describe the relation between terminal voltage and current when one varies in direct proportion to the other so that the graph representing this relation is a straight line.

Digitals are referred to as non-linear which implies that the ratio of voltage and current differs at different operating points and the output varies in a discrete (discontinuous) form.

Analogue circuits

While the variety of analogue ICs is well established only a few standard circuits stand out as having the widest application in systems of various kinds. These include operational amplifiers, voltage regulators and analogue-to-digital (ADC) and digital-to-analogue (DAC) converters. Today the operational-amplifier (op-amp) especially in the form of an IC has become the basic component for the analogue monolithic circuit. The low cost and high performance of the modern op-amp has rendered it almost universal in analogue circuit designs. Op-amps are often used in conjunction with discrete resistor–capacitor components and with thin or thick *RC* components in hybrid ICs to perform various signal filtering functions. Because of size and cost inductors are not used and hence useful values of inductance are not directly achievable in monolithic circuits.

Recently developed approaches in the design of analogue monolithic circuits are making possible the high densities usually associated with digital integrated circuits such as logic chips, memories and microprocessors. As a result, chip designers are combining traditional analogue functions, such as linear-amplifiers, ADCs and DACs and frequency filters, with digital functions on the same LSI chip.

Combinations of analogue and digital functions on a single chip have been achieved by application of complementary metal oxide semiconductor (CMOS), *n*-channel MOS and bipolar I²L techniques. One of the first major applications for monolithic ADCs and DACs has been in telephone systems using digital transmission and switching of voice signals. Here ADCs and DACs are basic to codecs (short for encoder-decoder). An encoder contains all the elements needed to convert analogue voice into digital data for transmission over a telephone line. Conversely a decoder contains all the elements that perform the digital-to-analogue conversion needed to restore the sound of voice after transmission.

In any telephone system the input voice information is an analogue time-varying signal. The signals continually vary in voltage or current, corresponding to the amplified variation of the input voice or data information. This tends to lead to accumulation of noise in the transmission. To avoid this the analogue signal is converted into a train of quantised pulses (a digital signal) before transmission (Figure 8.2). At the receiving end the binary signal is fed into a DAC, which restores the analogue signal. The transformation from the analogue signal to the digital signal is known as the coding of the signal. This modulation in accordance with a pulse code is known as Pulse Code Modulation (PCM) and has represented a major contribution to the communications industry.

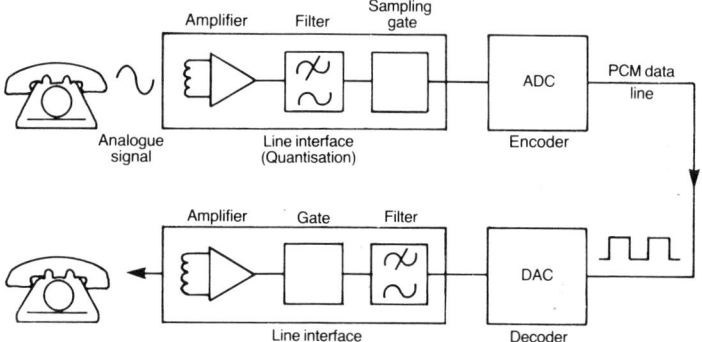

Figure 3.2. Basic PCM system involves encoding the analogue voice signal into a PCM signal. The resulting binary bit stream is decoded at the receiver for reconstruction of the original analogue pulse-modulated form

Operational amplifiers

Application of negative feedback to a high gain d.c. amplifier produces a circuit with a precise gain characteristic that depends on the feedback used. By the proper selection of feedback components, op-amps can be used to perform various mathematical operations. The term operational is indeed derived from the fact that the first operational amplifiers were used in analogue computing to perform addition, subtraction, averaging, integration and differentiation. They are now more widely used as the basic building blocks of electronic control systems and by using resistance–capacitance networks in the feedback, the overall characteristics allow the design of active filters and selective pass amplifiers. Op-amps generally function as two or more stage assemblies designed for insertion into other equipment such as analogue and digital computers, servo-systems and various types of feedback networks.

The majority of op-amps have a differential input but almost all have a single output terminal. Op-amps are generally represented by the triangular symbol shown in Figure 8.3 with two input terminals. A signal applied to input V_1 results in an output with inverted sense and is denoted as negative, whereas a signal applied to input V_2 results in an output of the same sense and hence is denoted as non-inverting and positive. The feedback circuit is always connected to

Operational amplifiers   147

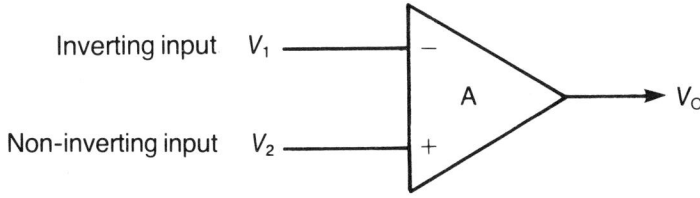

Figure 8.3. Operational amplifier (op-amp) symbol

the inverting input for negative feedback but the signal may be applied to either terminal depending upon the required input impedance of the amplifier.

The circuit arrangement of the amplifier unit which forms the basis of most IC op-amps is shown in Figure 8.4 and consists

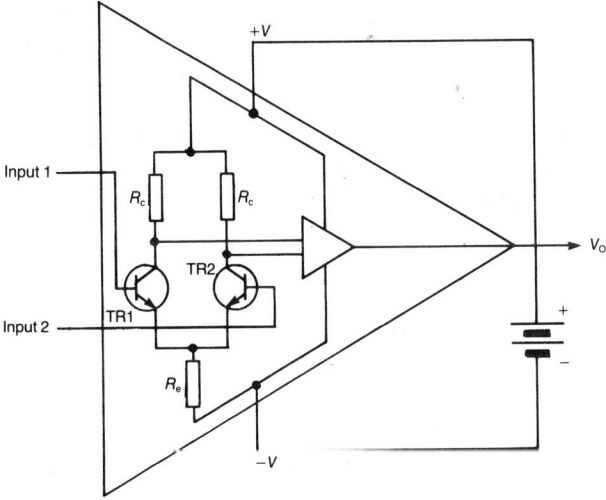

Figure 8.4. Structure of long-tailed pair of the op-amp. Although the amplifier is shown with bipolar transistors MOSTs are equally effective

essentially of two transistors, TR1 and TR2, with identical collector resistors, coupled by means of a common extended (tail) emitter resistor R_e across which the input signal for the second transistor (TR2) is developed. This basic circuit known as the 'long-tailed' pair is a differential amplifier in that it amplifies differences between its two input voltages but is not dependent on the magnitude of either input voltage. For instance if two equal signals are applied simultaneously to both inputs, each input voltage produces a voltage change at its own collector and an equal and opposite change at the other collector. Hence the output will be zero. On the other hand, if there is a difference in potential at the two input terminals the current through one transistor increases, while that of the other will fall, yielding an output that is an amplified version of the difference between the potential at the two input terminals. In other words the long-tailed pair produces an output that is proportional to the difference between the signals applied to the input terminals.

Common-mode

As mentioned above, like signals at the inputs yield zero output. This equal polarity signal is known as common mode and is a voltage that appears in common at both input terminals with respect to the output reference (usually ground potential). The quality of a differential amplifier may be measured in terms of its ability to amplify only the difference between the input signals and to reject the signal common to both inputs. A factor that expresses this degree of rejection is known as common-mode rejection and may be expressed as a ratio

$$\text{Common-mode rejection ratio} = \frac{\text{Gain with differential signals}}{\text{Gain with common-mode signals}}$$

The higher this ratio, the more efficient the amplifier. It is usually expressed in decibels, 100 dB being typical.

Drift

Any change in the supply voltage to the long-tailed pair will affect the transistor collectors equally and hence will not influence the signal being amplified. Similarly, temperature changes and internal bias variations in the two transistors will tend to be neutralised. This is an extremely important property for it means that drift is either rejected or reduced to a very low level. Drift may be defined as any undesired change in the output of a d.c. amplifier that is not due to a change in the input. The main causes of drift are changes in temperature and supply voltage which cause variations in the d.c. bias conditions. This results in spurious changes affecting the output.

Basic configuration

The basic arrangement for applying negative feedback voltage to the op-amp is shown in Figure 8.5 in which the feedback has been

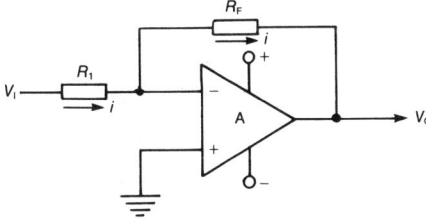

Figure 8.5. Basic arrangement of op-amp circuit

applied between the output (V_O) and input (V_I) of an amplifier (A) by means of resistors R_1 and R_F.

The current i through $R_1 = \dfrac{V_1}{R_1}$

and the output voltage $= V_O = iR_F$
Substituting

$$\frac{V_O}{V_I} = \frac{R_F}{R_1} = A_{vf}$$

Hence the feedback amplification is controlled by the ratio of the two resistances.

Application

As previously mentioned an important application of op-amps is to perform mathematical operations such as addition, subtraction, multiplication etc. To carry out these operations different combinations of passive components are used.

Addition

Figure 8.6 shows a circuit made to carry out the function of addition, consisting of three input signal voltages applied via three series resistors to the inverting terminal of the amplifier. Since all the input

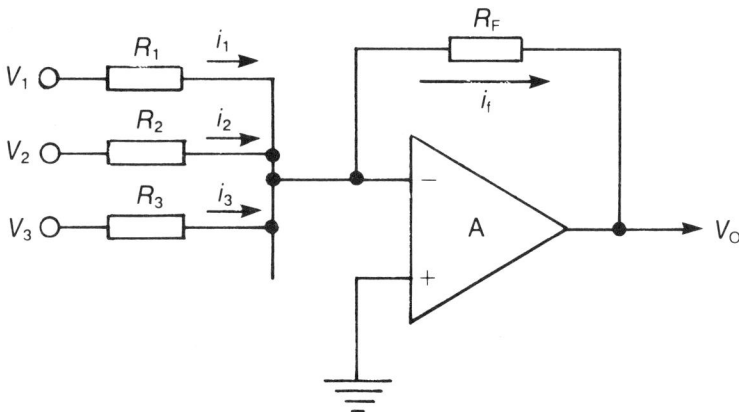

Figure 8.6. Summing op-amp inverting circuit

current flows through the feedback resistor R_F, the current is equal to the sum of the three input signal currents.

That is $i_1 + i_2 + i_3 = i_f$
Since

$$i_1 = \frac{V_1}{R_1}; \quad i_2 = \frac{V_2}{R_2}; \quad i_3 = \frac{V_3}{R_3} \text{ and } i_f = \frac{V_o}{R_F}$$

$$\frac{V_1}{R_1} + \frac{V_2}{R_2} + \frac{V_3}{R_3} = \frac{V_o}{R_F}$$

and

$$V_o = R_F \frac{V_1'}{R_1} + \frac{V_2'}{R_2} + \frac{V_3'}{R_3}$$

If all the resistances are made equal then the output voltage is directly proportional to the sum of the inputs.

Subtraction

Op-amps yield an output that is proportional to the difference between two input signals, that is the circuit functions as an inverting amplifier to one input and as a non-inverting amplifier to the other input. Consequently the outputs tend to oppose each other, the output being equal to input 1 minus input 2 and hence the circuit can be used to carry out the function of subtraction.

Integration

When the feedback resistor is replaced by a capacitor, the op-amp circuit becomes an integrator (Figure 8.7). When an input voltage V_1

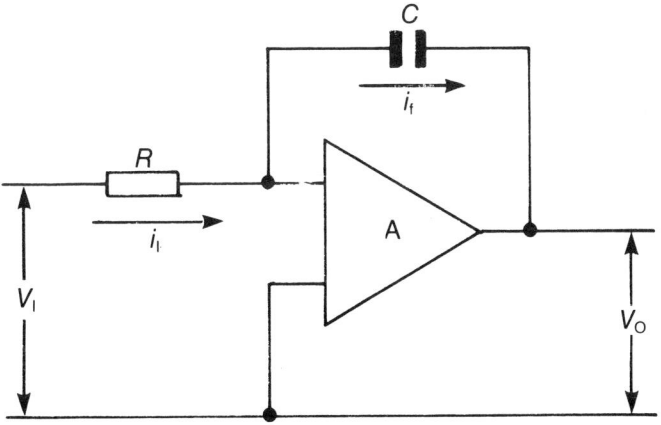

Figure 8.7. Op-amp integrator

is applied across the RC combination and an output V_O is taken across the capacitor, the current (i_1) set up by the input voltage charges the capacitor. The output voltage developed across the capacitor represents the accumulated charge which in effect is the integral (summation) of the charging rate. By making the charging rate a function of the input voltage, the capacitor voltage will represent the integral of the input voltage which is the quantity to be computed. In this case using instantaneous values, the charging current i_1 and the current i_f are the same and equal to V_1/R.

The current in a capacitor is proportional to the rate of change of voltage.

Hence

$$i_f = C\frac{dV_O}{dt}$$

Solving for V_O

$$V_O = \frac{1}{C}\int i_f dt$$

Substituting for the current ($i_f = \frac{V_1}{R}$)

$$V_O = \frac{1}{CR}\int V_1 dt$$

Hence the output voltage is equal to the integral of the input voltage multiplied by the constant $1/CR$ which is usually chosen as unity.

Differentiation

Differentiation is the inverse of integration hence by interchanging the capacitor and resistor (Figure 8.8), that is when a capacitor is

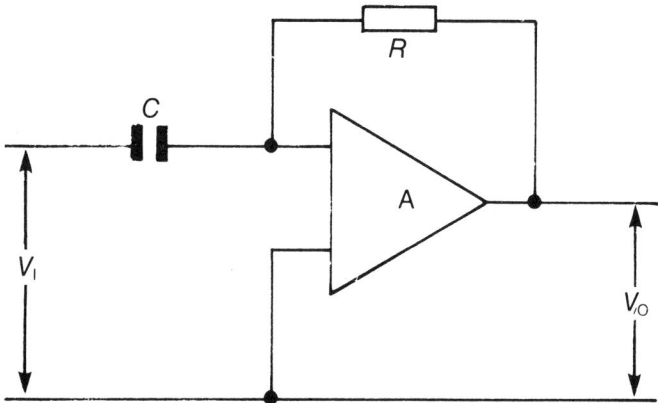

Figure 8.8. Op-amp differentiator

used to provide the input impedance and a resistor to provide the feedback impedance a differentiation circuit is obtained. By analysis similar to the above it can be shown that the op-amp will differentiate the input signal

$$V_O = RC\frac{dV_1}{dt}$$

Hence the output voltage is equal to the product of the constant RC and the rate of change of the input voltage with respect to time.

Voltage comparators

A voltage comparator as its name implies is a circuit used to compare the magnitude of two analogue signals and develops a logic output when the voltages being compared are equal or when one is greater or less than a reference level. When operating without feedback, op-amps make fast and accurate voltage comparators. The output is at a low voltage level (logic '0') when the input is more negative than the reference voltage; it is at a higher level (logic '1') whenever the input is more positive than the reference.

The basic component of all ADCs are analogue comparators which serve as output registers (Figure 8.9). This binary output

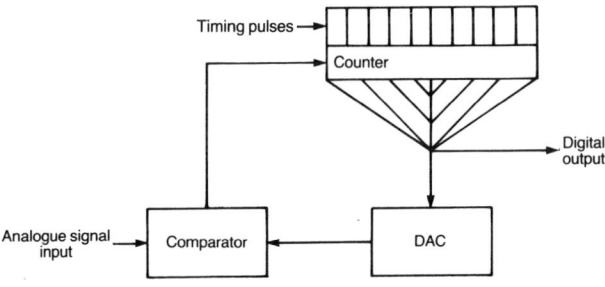

Figure 8.9. Analogue-to-digital converter (ADC)

indicates the sign of an analogue input sum or difference. As such a comparator serves to interface between analogue and digital systems.

Comparators may take many forms and find many uses. For example an electronically regulated d.c. voltage supply may use a circuit which compares the d.c. output voltage with a fixed reference level. The resulting difference signal controls an amplifier which in turn changes the output to the desired level. Comparators are also used to drive lamps, relays and MOS logic types.

Voltage regulation

The regulation of the voltage across a load for variation in either the load or the supply voltage is a critical functional requirement for the operation of most integrated circuits and solid-state circuit assemblies. Hence the primary function of any voltage regulation is to hold the voltage in its output circuit at a predetermined value over the expected range of load currents. Opposing factors are variations in load current, input voltage and temperature. A voltage regulator (Figure 8.10) uses an error or difference signal to correct any error (variation) in the output. The voltage difference between a reference and a portion of the regulated output is detected and amplified. The control circuit senses the magnitude of the amplified difference and regulates the load voltage in the appropriate direction to correct any voltage change. A stable reference voltage can be generated by the use of a Zener diode which is characterised by an almost constant voltage over its specified current range. A common type of voltage regulator is the 'series' regulator shown in Figure 8.10. The name 'series' comes from the fact that the output voltage is controlled by a

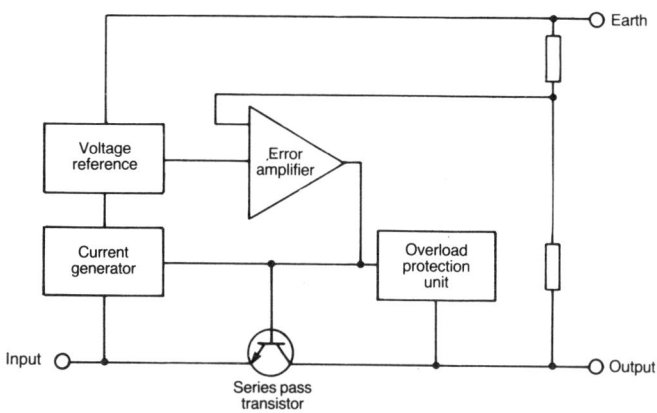

Figure 8.10. Block diagram of electronic voltage series regulator unit

power transistor in series with the output. Referring to Figure 8.11. TR1 is the series transistor, TR2 is the amplifier and D the Zener diode. Any increase in the output voltage V_o increases the voltage

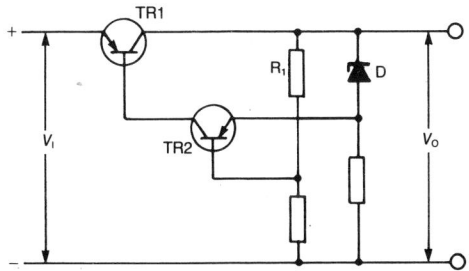

Figure 8.11. Transistor series stabiliser circuit

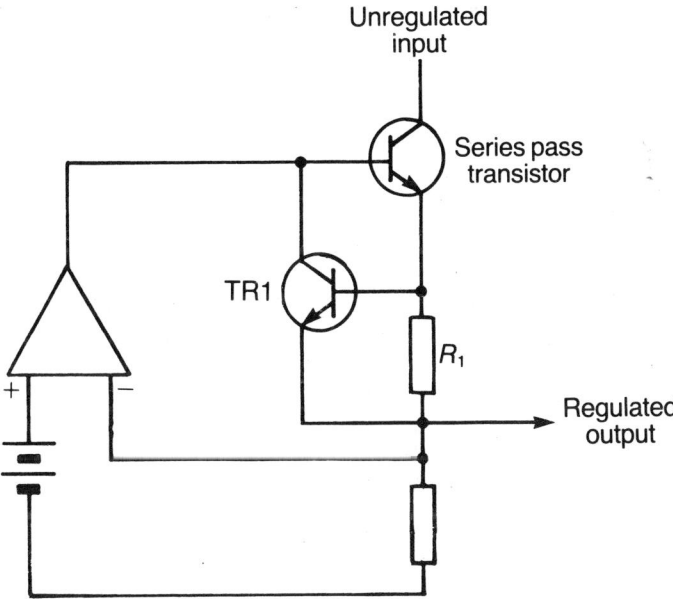

Figure 8.12. Analogue series regulator

from emitter to base, in effect increasing the voltage across the transistor TR2. This in turn increases the base current and the potential drop across TR1 which counters the original increase in V_o.

Voltage V_O is thus reduced, approaching its original value. Conversely, if input voltage V_I increases, the emitter-to-base voltage increases and the transistor voltage rises to overcome most of the increase. The Zener diode, biased by resistor R_1 into the required portion of its reverse characteristic, is used as a reference source. The regulation in this circuit is not perfect because small changes in output voltage occurs in the presence of a much larger change in input voltage.

Zener diodes' breakdown voltage occurs at a low reverse voltage — typically about 6 V — which puts a lower limit in the input voltage to the regulator. Substantial noise is also introduced into the circuit by the avalanching diode.

To circumvent these disadvantages an analogue circuit as shown in Figure 8.12 is used. An op-amp compares a reference voltage with a fraction of the output voltage and controls a series pass transistor to regulate the output. Overload protection is provided, the output current being limited by R_1 and TR1.

Impedance

An important consideration when choosing a circuit is the input and output impedance requirements, the reason being that it is a measure of the effective resistance in a circuit when the current flow is a.c. It may be defined as the ratio of voltage to current in an a.c. circuit and

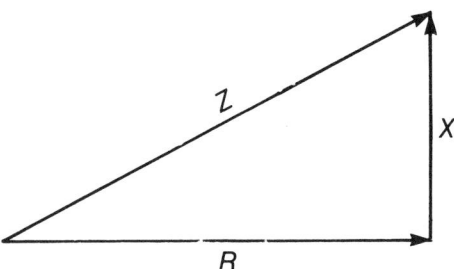

Figure 8.13. Impedance triangle. A diagram consisting of a right-angled triangle with sides proportional to the resistance (R) and reactance (X) of an a.c. circuit. The hypotenuse then represents the impedance (Z). The cosine of the angle between resistance and impedance is equal to the power factor of the circuit

hence can be considered as the a.c. form of Ohm's law. Thus $Z = V/I$, the impedance symbol being denoted by Z. Like resistance it is measured in ohms. It depends on the combined effect of inductance and capacitance known as reactance (X) as well as the resistance. Although resistances in series are simply added to obtain the total resistance, reactance and resistance must be added in a special manner to obtain the total impedance. Consider the right-angled triangle in Figure 8.13. The length of the sides adjacent to the right angle represent the ohmic values of the resistance (R) and reactance (X), the hypotenuse representing the impedance. By Pythagoras's theorem $Z = \sqrt{(R^2 + X^2)}$. For instance if $R = 80\ \Omega$ and $X = 80\ \Omega$, $Z = \sqrt{(80^2 + 80^2)} = 113\ \Omega$.

If reactance and resistance are in parallel, then

$$Z = \frac{RX}{R^2 + X^2}$$

Input Impedance is the ratio of the voltage applied to the input terminals to the current flowing into the input terminals and is represented by

$$Z_{IN} = \frac{V_{IN}}{I_{IN}}$$

Output impedance can be stated in terms of the unloaded output voltage V_O of the device and the loaded current I_L and voltage V_L.

$$Z_{OUT} = \frac{V_O - V_L}{I_L}$$

Hence the drop in output voltage $V_O - V_L$ as the device is loaded is determined directly by the output impedance.

The output impedance can be determined by connecting a voltage source to the output and varying the applied voltage by an amount ΔV hence varying the current introduced by the external voltage source into the circuit by an amount ΔI, the output impedance is then given by $\Delta V / \Delta I$.

Admittance

This term is encountered in transistor circuitry being denoted by the symbol Y. Admittance is the reciprocal of impedance. Thus if impedance equals V/I admittance equals I/V. For example, if impedance equals 10 Ω admittance will be 1/10 or 0.1 S, the unit being siemens. Calculations with admittance are exactly the reverse of those with impedance; impedances in series are added, admittances are added when they are in parallel. Summarising it may be said that all operations based on voltages with impedance, are based on currents with admittance.

Filters

Filters are in use in many types of electronic circuits. Radio and television receivers utilise filters to pass one particular channel and impede others on the basis of their different frequency bands. A telephone network incorporates literally millions of filters for the purpose of passing each conversation carried by a different band of frequencies to the appropriate telephone receiver and to stop all others. The components are known as filters because they function by separating signals of different frequencies, passing signals of the required frequency and rejecting or attenuating all others. The frequency bands in which there is negligible attenuation constitute passbands and those that are attenuated make up stopbands. The location in the frequency domain of the passband and stopband are used to classify the filter as low-pass, high-pass, bandpass or bandstop.

Filters are utilised (a) to resolve signals into their frequency components, (b) to eliminate signal contamination such as noise, rumble, hum etc., and (c) to remove signal distortion. They consist mainly of reactances which can take many physical forms such as resistors, coils and capacitors and in fact filters are classified into passive and active according to the nature of their components. Passive filters are constructed with capacitors, resistors and inductors, although for certain frequency ranges inductors because of their size and practical performance limitations are unsuitable. Inductor/capacitor filters, for example, are restricted to the approximate range 100 Hz to 300 MHz. Consequently, there has been a trend towards replacing inductors by active devices usually op-amps. This trend has accelerated with the advance in miniaturisation.

The important characteristics of a filter are given in Figure 8.14. The filter attenuation is given in terms of decibel drop at a given frequency and the passband is defined as the frequency region at which the response is attenuated to a drop of 3 dB. The point is termed the cutoff frequency of the filter and separates the passbands and stopbands and has the symbol f_c if there is only one cutoff as in low-pass and high-pass filters and f_1 and f_2 if there is more than one cutoff such as in bandpass and bandstop filters.

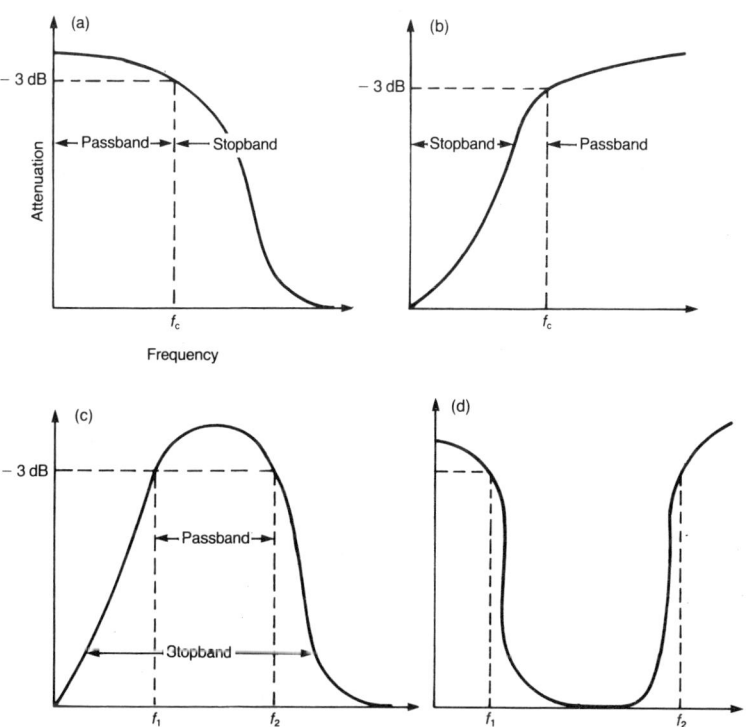

Figure 8.14. Characteristics of filters. (a) Low pass; (b) high pass; (c) bandpass; (d) bandstop

Basic filter operation

Filter operation is based on their offering high or low impedance to the different frequency bands. To see how a filter operates it is helpful to consider a basic type of filter (bandpass) as shown in

160 Analogue circuitry

Figure 8.15c. Assume an input signal has a frequency range of 1 MHz to 5 MHz and the requirements are for a 4-MHz signal with rejection of signals on either side. To function both the series and

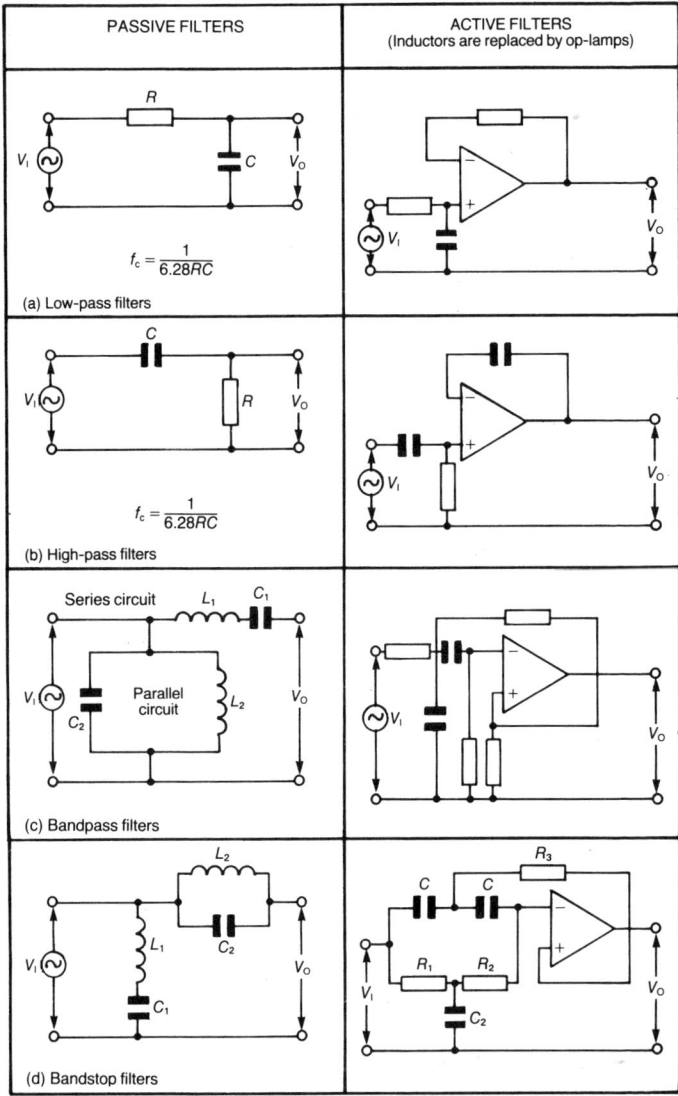

Figure 8.15. Passive and active filter circuits

parallel circuits are tuned to the required signal frequency. Thus for the required 4-MHz signal the series circuit offers a low impedance but offers a high impedance to currents of other frequencies and attenuates such signals. The parallel circuit on the other hand offers a high impedance to the 4-MHz signal and relatively little impedance to currents of other frequencies.

As a consequence the undesired currents are by-passed through the parallel circuit. By regulating the resistance and the L/C ratio the impedance for the desired and undesired signals can be regulated to meet specific requirements.

Types of filter

Low-pass filters pass frequency signals up to a specified cutoff frequency but attenuates high frequency signals above the cutoff frequency. In other words the passband extends from zero to some fixed frequency limit and the stopband extends from f_c to infinity. In the circuit shown in Figure 8.15a, at very low frequencies the capacitor C offers a very high impedance compared with the resistor R and nearly all the voltage applied across them appears across C. As the frequency rises the impedance of C falls until, at a very high frequency, most of the voltage appears across R. The dividing frequency (f_c) at which the voltages across C and R are equal is given by:

$$\text{Capacitance reactance } (X_c) = \frac{1}{2\pi f_c C} = R$$

Hence

$$f_c = \frac{1}{2\pi CR}$$

Scratch filters used in record players are low-pass filters designed to attenuate high-frequency noise.

Simpe RC filters are inadequate for many purposes because of the very gradual transition from the passband to the cutoff region, that is cutoff is not very sharp. A more rapid transition at cufoff may be obtained by using a two- or three-section RC filter. Because the response characteristics of a low-pass filter determine its effectiveness in separating signals, the low-pass filters are classified according to the manner in which the filter may affect the transmitted signal. They are known by the name of the individual credited with the first description of that particular circuit, for example Butterworth, Bessel, Chebyshev.

High-pass filters
This type of filter is derived from low-pass filters by the interchange of the two components, the resistor's being in series with the capacitor (Figure 8.15b). As a consequence the characteristics are reversed, frequencies below the cutoff frequency are attenuated whereas signals above the cutoff suffer no attenuation and are passed to the load.

This type of filter is commonly associated with record player circuits to eliminate 'rumble'. Rumble is low-frequency noise introduced by turn-table motors which induce spurious low frequency signals into adjacent electronic circuits. To overcome this, *RC* circuits are inserted in the preamplifier stage.

Bandpass filter (Figure 8.15c)
In this type of filter the stopband includes both high and low frequencies and contains a passband between the finite frequencies f_1 and f_2. Its operation has already been briefly considered on p. 160.

Bandstop filter
A filter having the inverse performance of that of a bandpass filter is a bandstop filter which stops (attenuates) a narrow band of frequency between the specified values but passes currents having frequencies outside this range. A basic type of bandstop filter is shown in Figure 15d. Both the series and parallel circuits are tuned to resonance at the centre of the band of signals to be eliminated. The parallel resonant circuit provides a high impedance whereas the series resonant circuit forms a low-impedance path. For signals above and below the resonant frequency the parallel circuit offers a low impedance but the series circuit furnishes a high impedance path. Thus currents of the resonant frequency will be stopped while currents of all other frequencies will pass through.

Filters are available in configurations other than analogue. For example a digital type filter acts on a sampled version of the signal to be filtered and may be realised through discrete logic hardware elements or by suitable programming of a digital computer which is fed with a sampled version of the digital filter. This type is, however, much more complex than an analogue filter and for this reason will not be examined here.

9 Digital logic technology

Digital pertains to digits and in digital technology as exemplified by the digital computer, processing is generated, transmitted and presented in the form of a sequence of digits.

Digital systems use standard electronic principles and use such components as transistors, diodes, resistors and capacitors. Integrated circuits play a very prominent role in digital logic operations and in fact it may be stated that the development of IC technology has been strongly motivated by the increasing demand on the performance and economy of logic circuits used in the digital computer. Conversely, the wide range of digital technology has only become a practical reality because of IC technology and its extension to large-scale integration (LSI). Digital circuits differ from conventional circuits in that the transistors are operated as switches that generally have only two states; on or off, conducting or non-conducting with no intermediate steps.

Binary system

Mathematical computation as found in a digital circuit is based on a system of counting with radix 2. In this system there are only two digits 0 and 1 termed binary digits abbreviated to 'bits'. It is manifestly simpler to manipulate numbers in the binary scale than in the decimal system for it is only necessary to distinguish two states as compared to the decimal system which is based on ten digits. As an illustration take the number 1110.

In the decimal scale this is equivalent to:
$(1 \times 10^3) + (1 \times 10^2) + (1 \times 10^1) + (0 \times 10^0) = 1110$
In the binary system it is equivalent to:
$(1 \times 2^3) + (1 \times 2^2) + (1 \times 2^1) + (0 \times 2^0)$
   8   4   2   0   = 14
Hence the binary equivalent of 1110 is 14

As a further illustration let us consider how the number 86 would be written in the two systems.

$(1 \times 2^6) + (0 \times 2^5) + (1 \times 2^4) + (0 \times 2^3) + (1 \times 2^2) + (1 \times 2^1) + (0 \times 2^0)$
Binary
  1   0   1   0   1   1   0
Decimal
  64 +   0   +   16 +   0   +   4   +   2   + 0 = 86

the binary equivalent being written 1010110.

Conversion from decimal to binary

Any decimal number may be converted to a binary number as follows:

(1) Divide the number by 2 and note down the remainder.
(2) Divide the quotient by 2 and note down the remainder.
(3) Repeat until there is no quotient.
(4) Write out the remainders in the *reverse order* to which they were determined. This is the required binary number.

For example the decimal number 20 is converted to its binary equivalent as follows:

$\frac{20}{2} = 10$ remainder 0

$\frac{10}{2} = 5$ remainder 0

$\frac{5}{2} = 2$ remainder 1

$$\frac{2}{2} = 1 \text{ remainder } 0$$

$$\frac{1}{2} = 0 \text{ remainder } 1$$

Hence the binary form of 20 = 10100.

This method may seem somewhat lengthy and cumbersome especially as the decimal system requires only two positions (units and tens) to represent 86 compared to the binary system's seven positions. It must be remembered, however, that the latter requires only two digits viz. 0 and 1 and further the computer operates at speeds measured in nanoseconds (thousand millionths of a second).

Logic circuits

These are so called since the relationship between their input and output voltages can be described by logic functions derived from Boolean algebra. This system of algebra named after the English mathematician George Boole (1815–64) is an algebra of logic as applied to switching in which only two states are utilised. A letter is allocated to each element in the same manner as in ordinary algebra, the two states being indicated by '0' and '1', for example A = '1' or A = '0'. To facilitate the solutions of problems the operation is broken down into three basic logic functions 'AND', 'OR' and 'NOT'. In the AND function designated by L = A.B.C etc., L is in logical state '1' if all the elements, that is A *and* B *and* C etc., are in state '1'. Thus A.B indicates A AND B and does not mean the arithmetical multiplication of one by the other. For the OR function L = A + B + C etc., L is in state '1' if one or more of the elements, that is A *or* B *or* C are in logical state '1'. The + sign is used as a shorthand for OR. The NOT function L = $\bar{A}$ indicates as inverse, that is L = 1 if A = '0' and L = '0' if A = '1'. To show such negation an overbar is used to express the letter symbols as having been inverted, $\bar{A}\bar{B}$. It is so named because the input and output logic levels are *NOT* the same under the conditions of operation.

The three basic functions are physically realised by binary logic gates, a gate being an electronic circuit which operates on two or more input signals to produce an output signal. It is called a gate since it can be either open to allow information to pass or closed to the flow of information.

The transistor as a switch

The logical operations of AND, OR and NOTE gates are commonly implemented by transistors which are operated as switches that generally have only two states, logic '0' and logic '1'. These correspond to the saturated and cutoff conditions, respectively.

In the switching circuit of Figure 9.1a if the input voltage is zero, no current flows and the collector voltage, that is the output voltage

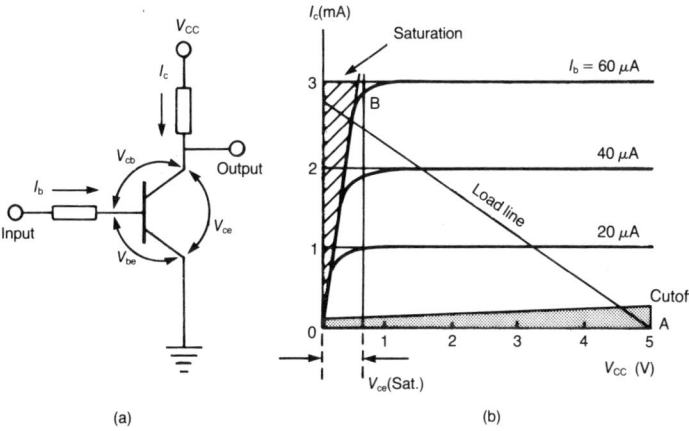

Figure 9.1. (a) Common-emitter transistor as binary switch; (B) transistor characteristic illustrating the logic 'l' and 'o' states

will be equal to the supply voltage V_{CC}, operation being at point A in Figure 9.1b which is the cutoff condition. In this state when the input signal is 0, the output will be a logic '1'. When, however, voltage is applied to the input terminal, forward bias of the emitter–base junction occurs causing an appreciable base current moving operation to point B, until eventually a point is reached when no further increase in collector current occurs and saturation ensues. At this point when the input is logic '1' the output is logic '0'.

Although a bipolar transistor has been considered in the above example many other semiconductors are used in logic gates and these will be considered in due course.

Logic functions

The three basic operations are indicated by symbols and their functions are defined by what are termed truth tables that give the desired output for every possible combination of units.

NOT gate

This is the simplest gate and is essentially a phase inverter and as indicated in the truth table yields a high output with a low input and vice versa.

The most common type of inverter is a common-emitter transistor (Figure 9.2). The supply voltage (a positive potential of about 5 V) is

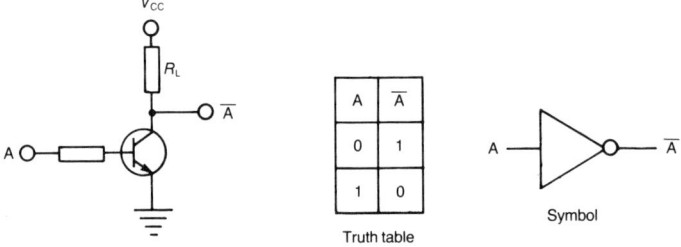

Figure 9.2. Transistor inverter NOT gate. This function inverts a binary digit changing 'o' to 'l' and vice versa

applied to the resistor R_L. The input signal is applied to the base of the transistor, the output being taken at the point where the collector resistor joins the transistor. When the signal at terminal A is in the low state, that is logic '0', the transistor is in the off condition and does not conduct and as a result the output is essentially connected to the supply voltage and is logic '1'. When a high signal is applied to the input the transistor conducts, the current produces a voltage drop through the resistor and hence the output is in the low state or logic '0'.

When the input is low, the output is high. If a high signal is applied the base of the transistor will be turned on, that is saturated and the voltage drop across the collector will be almost zero and can be considered to be a logic '0' output. The NOT gate is so named because the input and output logic levels are NOT the same under the conditions of operation. In the gate symbol a small circle at the output denotes that inversion occurs at the output.

AND gate

The AND gate has two diodes with their *p*-type region connected together (Figure 9.3) and as indicated in the truth table, an output

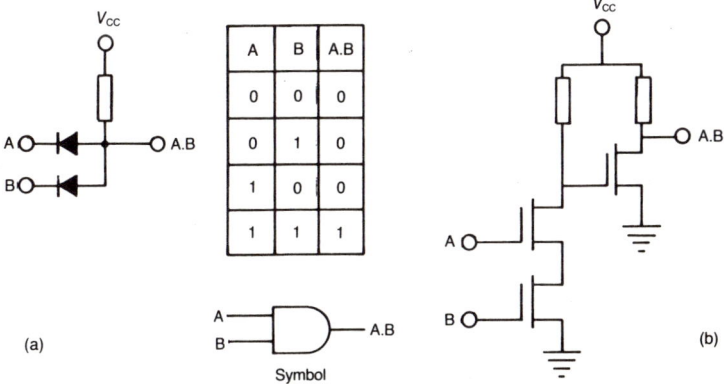

Figure 9.3. (a) Two-input AND gate; (b) MOST two-input AND gate. The AND function is a '1' only if both A and B are '1's

appears only when A and B are energised by a high signal simultaneously. In this state neither diode can conduct and the output rises to some voltage near that of the source voltage (V_{CC}). If any one or both inputs is zero, the corresponding diode is forward biased and therefore conducts to give a low output or logic '0'. In Figure 9.3b the gates are constructed from MOS transistors. The two input transistors are connected in series; current flows through them only when both receive a high signal simultaneously. To restore the proper polarity of the signal, the output of the two transistors is followed by an inverter.

OR gate

In this circuit the diodes have their *n*-type regions connected together (Figure 9.4). As indicated in the truth table a logic '1' applied to

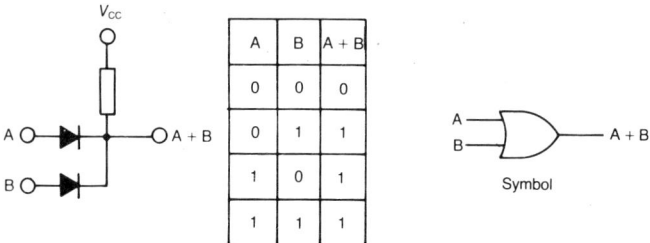

Figure 9.4. Two-input diode OR gate. This generates a '1' output if either A or B is '1' or if both inputs are '1's

either or both inputs causes a logic '1' to appear at the output. A logic '0' at both inputs retains the output at '0'.

Although for the sake of simplicity the number of different inputs applied to a gate in the above examples has been restricted to two any number of inputs may be applied simultaneously to a gate. The expression fan-in is used to designate the number of inputs.

Logic arrays

The AND and OR circuits in conjunction with the NOT type perform a variety of very important functions. The AND gate in combination with the NOT gate is known as a NAND gate (for not-and) and the NOT gate in combination with an OR gate yields a NOR circuit (for not-or). In fact logic circuits generally refer to NAND and NOR gates, since most digital circuits such as bistable circuits, storage elements, shift registers etc., can be constructed from these two basic gates.

NAND gate

The NAND gate is shown in Figure 9.5 from which it will be seen that the AND junction is represented by the diode AND gate followed by a transistor inverter switch (NOT gate). The NAND function is defined by the truth table. The AND gate performs the usual AND function producing a logic '1' only when both inputs are '1's. The NOT circuit inverts this intermediate AND signal to give the final output. Hence when any input to a NAND gate is energised by a logic

'0' signal, then its output is '1'. Otherwise the output is '0'. The NAND circuit also operates as an OR circuit at a low level, that is if the applied voltages are low in the desired state, then the output voltages will be high.

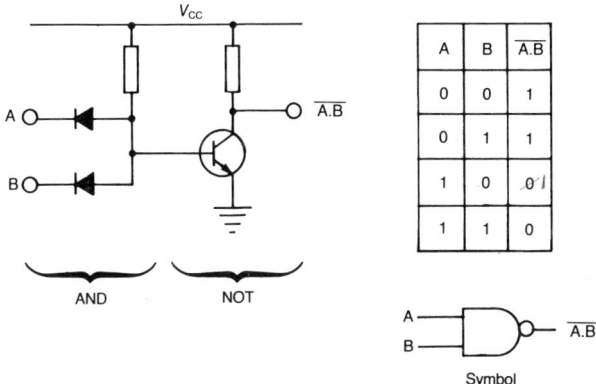

Figure 9.5. Two-input diode NAND gate. Can be regarded as an AND-NOT gate

The NAND circuit enables all the basic logic functions to be obtained, provided that the input signals are applied with the correct polarity. This leads to two most significant economic advantages. Firstly, fewer units are generally required to make up a given circuit than when other basic circuits are chosen. This means that fewer circuit blocks, printed wiring boards, connectors etc., are required. Secondly, because of the restriction to one kind of circuit, not only are fewer units needed to make a piece of equipment but also fewer different types are required. This simplifies manufacture and servicing and in fact it is quite possible to design the logic of an entire computer with nothing other than NAND circuits.

NOR gate

This is a combination of the OR and NOT gates (Figure 9.6). The OR gate performs the usual OR function producing a logic '1' whenever a '1' signal is applied to either input, its being inverted by the NOT gate. A NOT gate has two input switching devices and one output which is located between the load device R_L and a transistor NOT output stage. One NOR circuit is usually incorporated into an assembly of circuits (as in a computer processor) by connecting its output to the input of one or more similar circuits, and by connecting

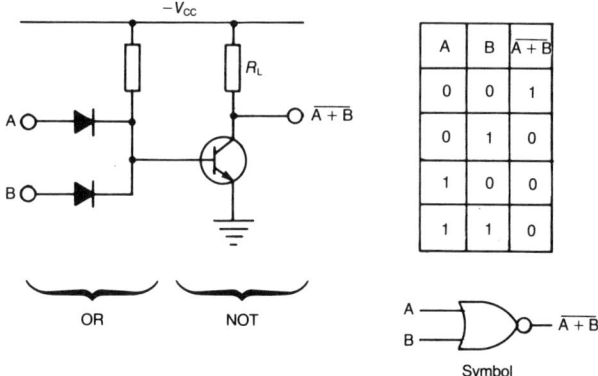

Figure 9.6. Two-input diode NOR gate. Can be regarded as an OR-NOT gate

each input to the output of one or more other circuits. If the voltage on any one of the inputs to the logic circuits is high (that is a binary 1) its switch will close and the output will be low corresponding to a binary 0. If all the inputs of the circuit are low (0) all the switching devices are open circuit and the output will be high (1). As in the case of NAND circuits it is quite possible to design the logic of an entire computer with nothing other than NOR circuits.

MOS logic

MOS transistors are widely used in logic circuits. The simple structure and low dissipation of MOS transistors allow a higher packing density than bipolar devices. One disadvantage is that the propagation delay time of p- and n-channel transistors is more than bipolar transistors which makes them slower in operation. This, however, can be remedied by the use of complementary MOS (CMOS) devices. CMOS devices also have several other advantages, one of the most important being the low current level which means low heat dissipation. This enables static logic to be produced that would encounter considerable problems of heat dissipation if made with n- or p-channel MOS transistors. Other advantages of CMOS compared with ordinary MOS circuits are their immunity to fluctuations in the supply voltage or in the input voltage. The sensitivity to input voltage fluctuation is low because the input voltage at which the circuit changes over from one logic state to the other is equal to about half the supply voltage while the actual transition takes place over a very small range of input voltage. CMOS gates have a high input impedance which is important in logic applications as it means that one stage is capable of driving several similar stages in parallel, that is there is a high fan-out. Basic to the use of CMOS in logic circuitry is the complementary (NOT) inverter gate (Figure 9.7). This is a circuit arrangement that complements a

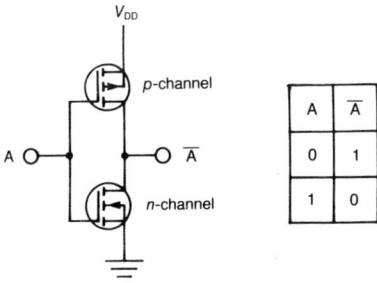

Figure 9.7. Inverter (NOT) circuit consisting of CMOS transistors

logic variable, that is gives a high output (logic '1') when a low or logic '0' is applied and gives a low output when a high input is applied. In this device the two gates are connected together as are the two drains. With a positive supply voltage V_{DD}, if a positive voltage A is applied to the input (logic state '1') then the n-channel transistor conducts while the p-channel transistor does not. The output voltage is then zero (logic '0'). If the voltage is now removed from the input ('0'), the p-channel transistor becomes conducting and the n-channel transistor is switched off. The output voltage is now V_{DD} ('1'). The two CMOS transistors therefore behave as switches and in both states the only current in the circuit is the leakage current of one transistor and hence dissipation is very low.

NOR gate

Figure 9.8 shows a NOR gate made from four CMOS transistors. The *n*-channel devices are connected in parallel and the *p*-channel devices in series. The gates of TR1 and TR3 are connected together

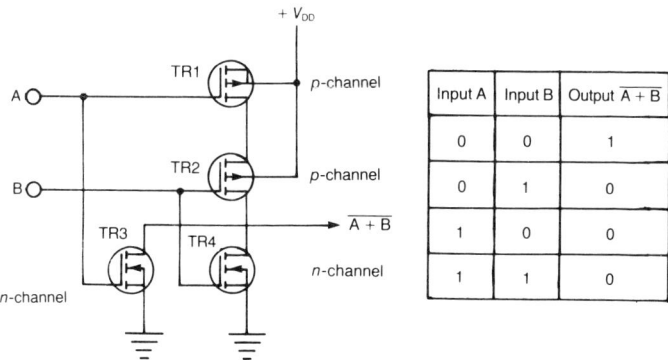

Figure 9.8. NOR gate consisting of four CMOS transistors

to form input A and act as an inverter circuit. Likewise TR3 and TR4 act as an inverter. With both A and B at a low level both *p*-channel transistors are conducting and the *n*-channel transistors are cut off, hence the output is logic '1' only when both inputs are at logic '0'. With a high-level input at either A or B or both, the *p*-channel devices are turned off and one or both of the *n*-channel transistors turn on. The result in each case is a logic '0' at the output.

NAND gate

In this gate (Figure 9.9) the *p*- and *n*-channel circuits are opposite to those in the NOR gate, the *n*-channel devices TR3 and TR4 are in series and the *p*-channel transistors TR1 and TR2 are in parallel. With both inputs at a low level, both *p*-channel transistors are conducting the *n*-channel devices being cut off. In this condition the output level is V_{DD} or logic '1'. Even with only one of the inputs at a low level, one of the *n*-channels is cut off and the output remains high. Both inputs are required to be at a high level for both *p*-channels to be cut off and both *n*-channel transistors to be conducting. The output voltage is then zero (logic '0').

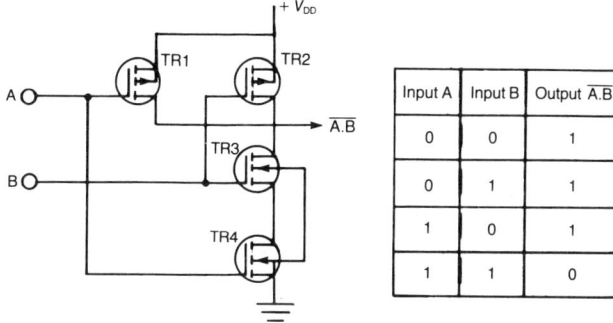

Figure 9.9. NAND gate consisting of four CMOS transistors

Integrated injection logic

A recent approach to LSI is Integrated Injection Logic (I²L) also known as Merged Transistor Logic (MTL). The I²L technique represents an important development in bipolar transistor circuits. It operates faster than *n*-channel MOS and consumes less power than CMOS which are, respectively, the high-speed and low-power branches of the MOS family. In the bipolar group it offers the means to achieve low power dissipation, high switching speeds and high packing density.

In I²L ICs vertical *npn* transistors each with a number of collector regions at the top and a common (grounded) emitter at the bottom are combined with a lateral *pnp* transistor which acts as a current generator, the emitter of which is referred to as an injector. On an IC the *pnp* injector is integrated or merged with that of the *npn* transistor in the base silicon — hence the alternative name.

The fundamental I²L circuit is an inverter consisting of a vertical *npn* multiemitter transistor operated in the inverse mode. A common (grounded) base *pnp* transistor supplies current to the base of the common emitter *npn*, the emitter of the *pnp* being the injector. In effect, injection of minority carriers occurs into the base region of the *npn* transistor by the lateral *pnp* transistor which is an integral part of the *npn* structure.

Figure 9.10a shows the cross-sectional view of the *pnp* and *npn* structure with a n^+-type substrate which acts as the common emitter for all *npn* transistors and as the base for all *pnp* transistors. Also represented is the circuit configuration. Figure 9.10b shows the interconnection of I²L to form NAND logic.

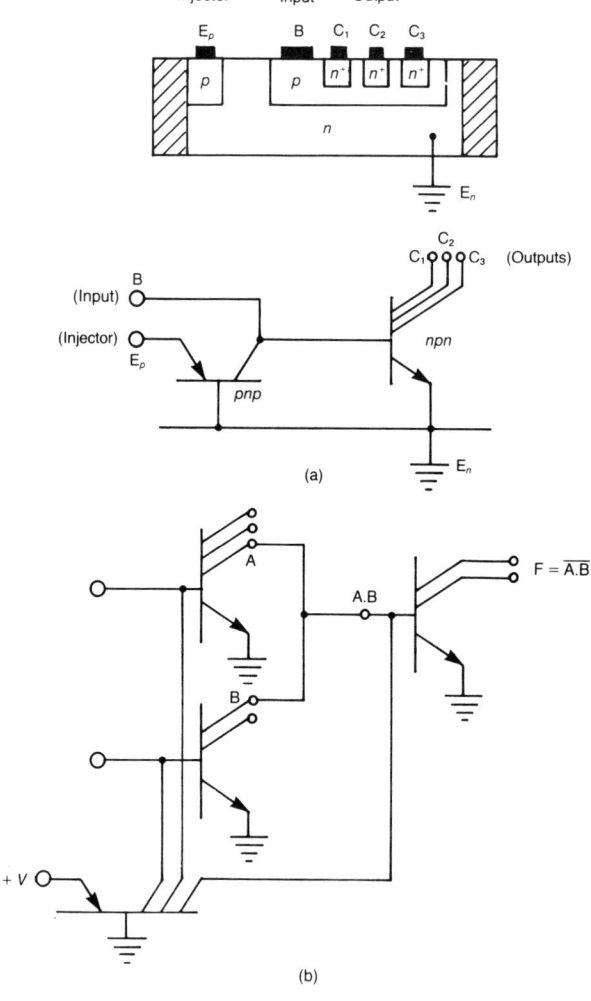

Figure 9.10. (a) I²L cross section and circuit representation. The lateral pnp transistor (emitter E_p, base E_n, and collector B) functions as a current source supplying base current for the operation of the inverted npn transistors (emitter E_n, base B and multiple collectors C_1, C_2, C_3. (b) I²L circuit connect to form NAND logic

I²Ls are replacing ECL and Schottky TTL which are currently the fastest operating digital ICs available. The high packing density, excellent power saving performance and functional versatility have led to their utilisation in such diverse applications as electronic watches, microprocessors, camera controls and many other consumer products.

Bistable (flip-flop) circuits

In addition to the logic gates so far considered a further important basic circuit called the flip-flop is used in many digital systems. In the logic circuits so far considered, the output signal is determined by the input signal levels. However, there are other circuits whose output signal depends not only on the input signal but also on feedback signals applied previously to the input. Such a circuit is known as bistable and a characteristic property is that its switching state does not change when the input signal disappears but remains the same until the next input signal switches it to the other state. As the name implies the two states of the bistable are both stable and either can be preserved unchanged for a period of time.

The stable state in which the circuit resides depends on the past history of the input signals. In other words the circuit has memory, that is it can store logic information in the form of '1's and '0's. This property is of great importance in digital systems in which the bistable circuit is widely used as a data memory store in computers, where large numbers of such circuits are connected in series to form registers. The bistable circuit is commonly known as a flip-flop. This curious name is derived from the fact that a pulse 'flips' the circuit into the unstable state, the circuit afterwards 'flops' back into the stable state. The basic flip-flop (Figure 9.11) consists of two input lines designated as the set line (S) and the reset line (R) and two output lines which can be placed in various '1' and '0' combinations by the inputs. Basically one output is '1' when the other is a '0'. One output is called the Q output or set output while the other is the $\overline{Q}$ or reset output (the complement of Q). If $Q = 1$ and $\overline{Q} = 0$ the flip-flop is said to be set or in the '1' state while for the reverse the flip-flop is reset or in the '0' state. Hence the two outputs are always in the opposite state, both being changed over by an input signal. The truth

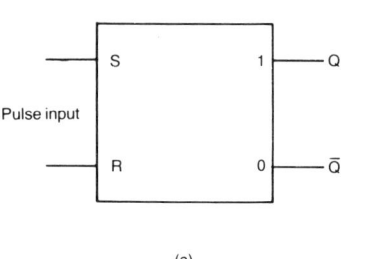

Figure 9.11. (a) S-R flip-flop logic symbols; (b) truth table

table for an S-R flip-flop is shown in Figure 9.11b. In Figure 9.11a the set output Q results from the '1' output terminal, and the $\overline{Q}$ from the '0' output terminal. When both inputs are at logic '0', there will be no change in state (Q_n). A logic '0' in the S and a logic '1' in the R results in a logic '0' whereas a '1' in the S and a '0' in the R results in the logic '1' or set state. If a logic '1' signal is applied to both S and R inputs simultaneously the result is an undetermined (U) condition, that is it may go to either the '1' or '0' state.

Flip-flop circuitry is composed of NAND gates or other circuit modules, for example NOR or NOT gates, which can be brought into one of two possible stable states by appropriate signal inputs to the S or R terminals. A basic bistable circuit consists of two NAND gates cross coupled as shown in Figure 9.12. Such a circuit is

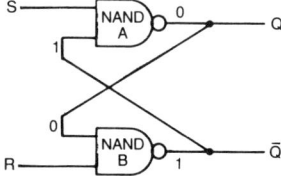

Figure 9.12. S-R flip-flop implemented with NAND gates

completely symmetrical, the output of one gate is connected to the input of the opposite gate. To understand the operation assume that gate A is conducting so that its output is logic '0'. This output constitutes the input to gate B. Since B inverts it is cut off and its output is logic '1' which goes to the input of gate A. Since A is conducting the situation is stable. Hence it can be seen that if both inputs to each gate are held in the logic '1' state then the '0' output of one gate will ensure that the output of the other gate is '1'. Always one gate is '0' and the other '1'. In other words, it provides outputs by maintaining complementary logic levels at its two output terminals. The flip-flop hence fulfils the requisites of a binary storage circuit, storing information by remaining in either the '1' or the '0' state until a new signal is applied to either of the Set or Reset terminals.

Digital logic ICs

The foregoing descriptions of various gate circuits form standard blocks for specific logic systems and generate a wide range of logical functions which are the basis of computer operation. The main difference between the various types is in the individual hardware components used in the circuits. The groups may broadly be classified by the mode of operation of the amplifier inverter transistors. One group uses the saturated mode for the transistor while the other group operates in the non-saturated mode.

Saturated mode

Resistor–Transistor logic (RTL)
This was one of the first type of digital logic gates to be produced and is also one of the simplest (Figure 9.13). The circuits are designed so

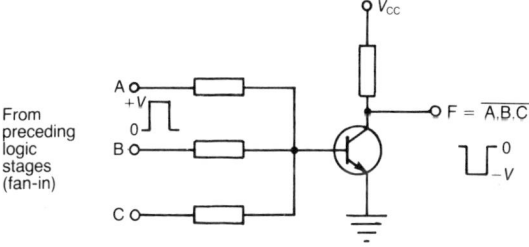

Figure 9.13. Resistor–transistor logic (RTL) gate

that logic '0' is represented by any voltage level less than 0.4 V. Logic '1' is represented by anything in excess of 0.7 V. The 0.3 V difference is intended to prevent noise interference. Each input is connected through a resistor to the base of the transistor. When any or all of the inputs is at a positive voltage, that is a '1', the base-emitter function is forward biased and the collector-emitter voltage is reduced to approximately zero giving a logic '0'. This is the equivalent of a NOT function. The values of the resistors are so chosen that the basic NOT circuit is large enough to saturate the transistor. The attractive features of the RTL are its simplicity and the use of relatively inexpensive components. Because resistors are used as coupling elements large signal swings can be used and this also makes possible a large number of fan-ins. In spite of this RTLs play little part today in logic systems chiefly because of a large propogation delay time and hence low speed.

Diode-transistor logic (DTL)

This circuit consists of a three-input diode gate (Figure 9.14) connected to a transistor inverter to establish the logic '1' and '0'

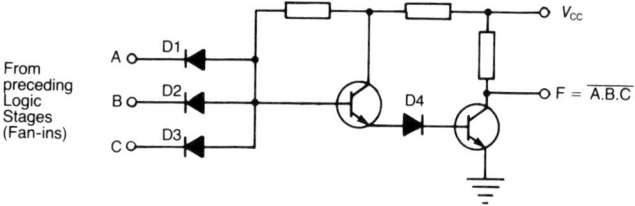

Figure 9.14. Diode-transistor (positive NOR gate) logic (DTL) gate

levels. The output of the diode gate is an AND function and the logic function generated at the transistor output is a NOT function. Hence this circuit is a NAND circuit. If the inputs to the diodes D1 D2 D3 are at logic '0' the diodes are forward biased and there is not sufficient current to pass through the diode D4 and the base-emitter junction of the transistor. Hence the transistor is cut off and the output is practically equal to V_{CC} to give a logic '1' output. When the input signal is switched to the logic '1' state, current flows into the base of the transistor and is large enough to saturate the transistor, the low collector voltage giving a logic '0' at the output. This type of circuit is very popular. Noise immunity, propagation speed and power dissipation are all good. Limitation on its fan-in capability is its principal shortcoming.

Transistor–Transistor Logic (TTL)

This family of bipolar integrated circuits is one of the most widely used today. It is unique in that it is based on a logic element consisting of a multiemitter transistor in which two or more emitters share a common base and collector and an output stage formed by one or a number of transistors (Figure 9.15a).

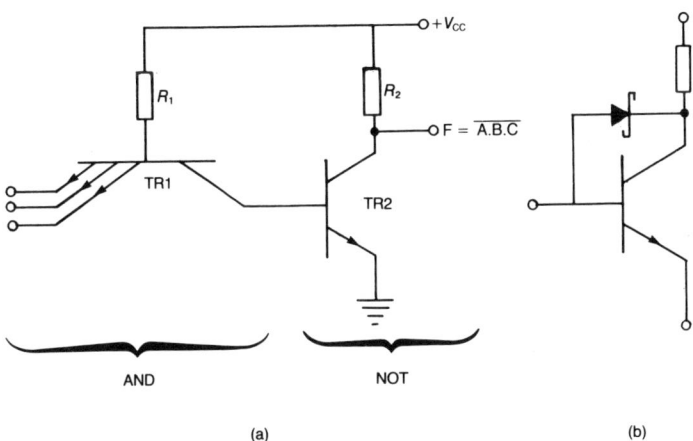

Figure 9.15. (a) Simplified TTL NAND gate. The circuit comprises a multiple-emitter transistor in which three emitters share a common base and collector and function as an AND gate. The remainder of the circuit operates as an inverter. (b) Schottky diode clamp connected from base to collector of transistor. Since majority carriers predominate there is no storage of minority carriers to limit switching speeds

If one or more inputs A, B or C are at logic '0', the emitter–base junction at TR1 is forward biased preventing the flow of current to the base of TR2 which is cut off. Current now flows through R_2 and a logic '1' appears at the output. When all the inputs are at logic '1' the emitters at TR1 are reverse biased, the potential of the base rises allowing sufficient current to flow through the base–collector junction of TR1 and drive TR2 into saturation to give a logic '0' output.

TTL has become a popular range in ICs and microprocessors are designed so that their signal voltages and power supply requirements are compatible with those of TTL circuits.

182 Digital logic technology

Schottky barrier diode TTL

As a result many variations exist to reduce power dissipation, propagation delays and increase noise margins. For example, when the transistors in the TTL circuit are driven into saturation to give logic '0' output, a significant amount of charge is stored as minority carriers in the base and collector regions and a relatively long delay occurs until they are discharged and the transistors are removed from saturation. Speed can be improved by incorporating a Schottky barrier diode (SBD) across the base-collector junctions (Figure 9.15b). Known as a clamp the SBD diverts most of the excess base current and prevents the transistors from reaching saturation. The feature of the SBD which makes it so useful is that it is free from minority carriers and therefore has no stored charge. With no storage in either the SBD or the transistors there is a great reduction in storage time which significantly improves transistor switching time.

Non-saturated mode

Emitter-coupled logic (ECL)

In logic circuits in which the transistor operates in the saturated mode, the time taken for the transistor to come out of saturation is a

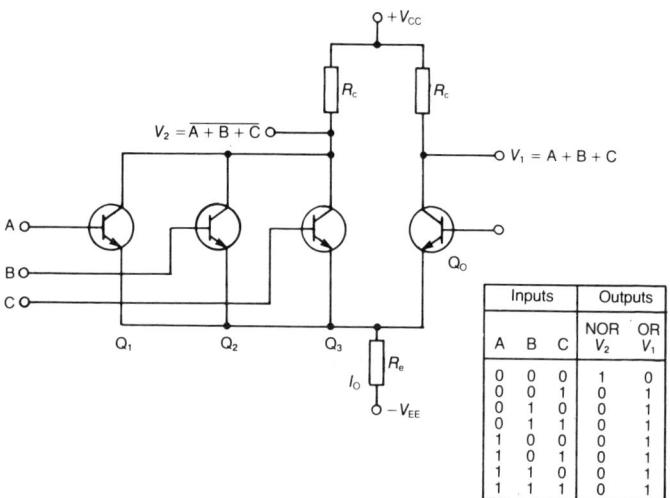

Figure 9.16. Emitter-coupled logic (ECL) gate

significant factor in signal propagation delay. Hence by operating the transistor in a non-saturated mode a resultant reduction in propagation delay is achieved and the circuit operates at high speeds. (In the case of TTL this may be accomplished by the use of a Schottky diode.) In ECLs the transistors are switched between two well-defined levels in such a manner that they never saturate, instead the transistor remains in the active or current-mode operation. The circuit consists of emitter-coupled transistors (Figure 9.16) hence the term emitter-coupled logic (ECL). It has the unique feature that two outputs are available, one of which is always the complement of the other. Both OR and NOR functions are available. The voltage $-V_{EE}$ and the emitter resistance R_e are such that they approximate a current source I_O. If negative signals are applied at all the inputs A, B and C then Q_1, Q_2 and Q_3 are cut off and the current I_O will flow through Q_O. Under this condition $V_1 = $ '0' and $V_2 = $ '1'. If a positive signal is applied to one or more of the inputs, the corresponding transistor(s) will turn on and the current I_O will no longer flow through Q_O. Hence $V_1 = $ '1' and $V_2 = $ '0'.

ECL is currently the fastest logic circuit, although speed is gained at a cost of high power consumption resultng in high heat dissipation within the IC. It is used mainly in large computers where its disadvantages can be suffered for the sake of high speed.

Logic parameters

In logic circuits the term 'noise' refers to any spurious signals that are not part of the input. The ability of digital logic ICs to reject the effect of these spurious pulses, for example from supply lines, is known as the noise margin or noise immunity. Together with the propagation delay these are the most critical of logic IC parameters.

The product of delay time and power dissipation per logic operation is known as the speed–power product. The smaller this number the better the gate. It is fairly consistant for a particular operation and hence it is sometimes used as a quality rating expressed in picojoules (1 pJ = 10^{12} watt-seconds). A typical value for good quality logic gates is roughly of the order of 100 pJ. Typical values of the more important parameters for the basic logic IC circuits are given in Table 9.1.

Table 9.1 Comparison of basic logic circuits

	RTL	DTL	TTL	ECL	TTL Schottky
Supply voltage (V_{CC})	3	5	5	3	5
Logic '1' voltage	0.7	2	3	0.4	3
Logic '0' voltage	0.4	0.4	0.2	−0.4	0.2
Noise margin (V)	0.4	1.0	0.7	0.4	0.7
Propagation delay (ns)	45	20	10	1	2
Power dissipation (mW)	10	10	10	40	10
Speed–power product (pJ)	450	200	100		20

10 Digital computers

This aspect of logic circuitry could well fill a large volume and space limitations in this book preclude describing all but the main areas of the subject.

Before discussing computers it is as well to answer the question 'What is a computer'. The full definition is very wide ranging, but in brief it may be stated that it is an electronic machine in which data are represented by electric signals, the machine following a program of instructions detailing the sequence of operations to which the data must be subjected in order to obtain an answer. In a basic sense a computer is a machine for transforming one set of data into another and hence is a data transforming device and in fact may be called a data processing machine. Data can be represented in either analogue or digital form, the difference between them being the methods used to represent numbers. Analogue computers represent information not by discrete numbers but by analogous physical quantities with continuously varying amplitudes such as voltages, currents or electrical wave forms. For example, a car speedometer uses the continuous displacement of a pointer over a scale to indicate the speed. This contrasts with digital representation which uses separate discontinuous steps for each of the digits which together express a numerical value as in a car's odometer for indicating distances travelled. Analogue computers find their principal applications in design engineering but suffer because their computing units are accurate only over a very limited range. In digital computers precision can be increased to any required degree by the use of more digits, 10 to 12 being commonly used. For this reason and also because there is less demand on the stability of electronic circuits, digital computers have become almost universal in a wide variety of industries and activities. In fact more than 90% of the computers in use today are of the digital type.

Hardware

Combinations and arrays of the various types of gate circuits make up digital equipment. It is the systematic arrangement, interconnection and sequential operation of these gates that constitute the electronic digital computer.

A simplified layout of a digital computer system is illustrated in Figure 10.1 from which one can see that it consists of three main

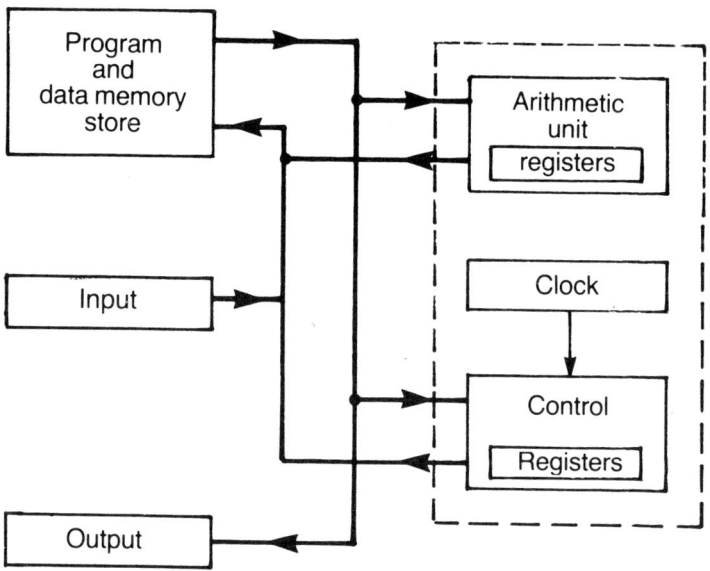

Figure 10.1. Simplified block diagram of computing system. The dotted square indicates the central processing unit (CPU)

parts, input/output ports, a memory or store and a central processing unit (CPU) comprising an arithmetic unit and a control unit. The CPU typically occupies three or four large steel cabinets.

The program or set of instructions which constitute the software are fed into the computer and take the form of punched cards, punched paper-tape, teletypes or transducers. These are converted into electrical impulses that represent data and are stored in the memory and are read in sequence by the CPU which carries out each instruction as it is received from the store. The arithmetic and logic unit (ALU) performs the necessary arithmetical operations. To extract the answers from the machine there is an output device which accepts data from the store and either prints it by means of electric typewriters or line-printers or displays it visually on a cathode ray tube.

Arithmetic and logic unit

The ALU is the section which actually performs the computation. In general it consists of registers (Figure 10.2) each consisting of a row

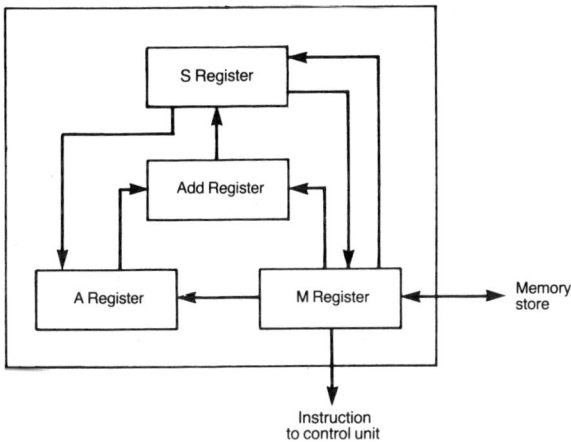

Figure 10.2. Arithmetic and logic unit

of flip-flops whose function it is to temporarily store the data during an operation, for example the multiplicand, multiplier and product during a multiplication. The register M forms the connection with the memory or store. When a number is brought from the memory to the ALU the memory supplies the requisite voltage pulses to set the flip-flops of M in the corresponding states.

The second register of the ALU is the accumulator A which may be regarded as the main working register into which data and results are written and processed and from which they are subsequently dispatched to the store or output port.

The third register S, known as the shift register, is used in multiplication and division, and consists of a number of cascaded flip-flops. The central part of the ALU is the adder (Add) which computates the sum of the numbers standing in A and M. During addition the sum thus formed is taken up in A and the previous contents of this register are cancelled out. Subtraction does not differ much from addition, the number transferred to M from the store is given an opposite sign and then added to the number in A.

Control unit

The control unit is concerned with the movement of data within the computer ensuring that the different operations take place in the proper sequence. The order in which the unit carries out the operations is set out as a series of instructions known as a program which is held in the store. The unit draws the instructions from the store, decodes them and delivers control signals which cause the appropriate action to be carried out. There are two aspects to the control function:

(1) Arithmetic control concerned with the ALU which needs control signals to perform each of the various operations as directed by the instructions. Some of these signals are generated within the ALU itself as the operation is being performed, others are generated externally.

(2) Instruction sequencing control is concerned with control signals that must be generated to direct the automatic execution and sequencing of instructions so that a complete program can be computed.

The processing of an instruction is controlled by a number of registers (Figure 10.3) one of the most important being the sequencer. This controls not only the arithmetic operation but also delivers the signals controlling the various stages of the 'fetch and execution' phases concerned in the processing of an instruction. For

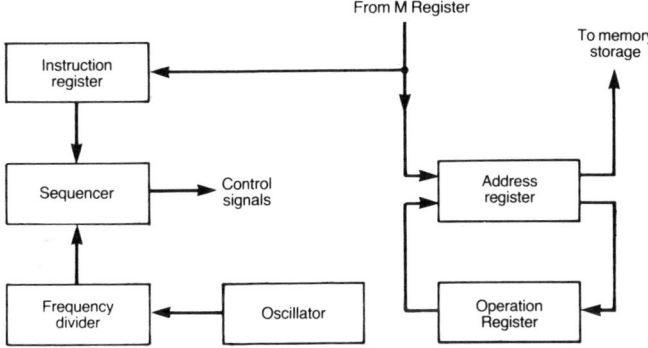

Figure 10.3. Simplified diagram of control unit

this purpose the sequence circuit contains a clock counter which receives pulses from a check pulse generator. This consists of a quartz crystal oscillator operating at a frequency of 6–8 MHz. The pulses, which are emitted at a predetermined rate and interval, are applied to the gates for the purpose of opening or closing them thus permitting transfer between the registers.

Memory (store)

The memory or storage is an essential function in all computer systems for it contains both the data needed for calculations and the instructions which control the operation of the machine. The registers in the ALU store the data during an operation, for example the multiplicand, multiplier and product during a multiplication, while the registers in the control unit store control information such as an instruction which is being executed. Information transfers therefore take place between the store and registers. Information is usually stored in 'words' placed in locations each of which is uniquely identified by a numerical address. When this address is presented to the access circuits, the information is either read or written into the corresponding location. It should be noted that the major problem in storage technology is not the actual storage of information but the means of getting to (accessing) the information when required. In 'serial-access' systems the stored words are presented sequentially at the output so that the time taken to access a location depends on its address. When the access time is independent of the address the system is known as 'random access' since the bits or words stored may be located and abstracted without affecting other data. Random-access memories (RAMs) consist of arrays of flip-flops which are set or cleared according to the data stored.

Another form of storage is the read-only memory (ROM) which differs from the RAM in that it stores data which cannot normally be altered. The memory contains a fixed pattern of '1's and '0's usually inserted during manufacture as specified by the customer. It can be compared to a dictionary. In Figure 10.4 the types of store required

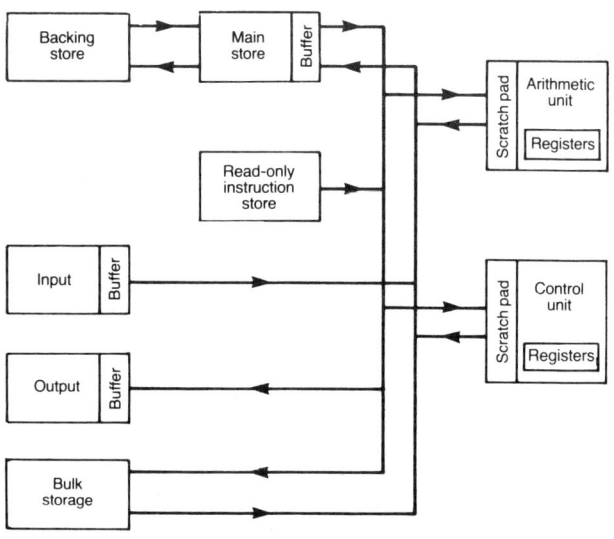

Figure 10.4. Types of store in a digital computer

in a modern computer are shown ranging from the high-speed registers where the capacity is several words only and the access time is of the order of a few nanoseconds, to the bulk storage which could be high speed, high density magnetic tape for example, where the capacity of one tape unit is over 10^7 words but with an average access time of 100 ns.

'Scratch-pad' memories are extensions of register storage where more registers are provided at lower cost by organizing them on a small storage system.

The use of a buffer store between the main store and the CPU allows overlapping in the operation of the store and the CPU. Buffer stores associated with the input and output peripherals simplify the transfer of information to and from the computer. The unit of memory is the 'bit' representing the binary 0 or 1 used for expressing information in a computer. A group of bits is designated as a 'byte' usually eight or nine bits or a 'word' 12-64 bits depending on the particular system.

Solid-state memories

In the past computer memories were almost without exception of the magnetic ferrite core type, composed of thousands of minute doughnut-shaped rings threaded together on a rectangular grid of wires. (This is why early 'main frame' computers were so large.) These small rings can be magnetised in either a clockwise or anticlockwise direction, the alternate directions being used to represent binary '0' and '1'. Today, developments in semiconductor technology have been applied to the high-speed storage field, and have led to the development of mini- and microcomputers. In this area MOS technology has made a considerable impact. MOS memory is about ten times faster than the magnetic core, uses less power and possesses a high packing density. Its stablemate, the bipolar semiconductor memory although more expensive is around 100 times faster than the core store.

MOS devices are self isolating and hence can be packed together much closer and thus take up less silicon chip area than is the case with bipolar devices. Memory circuits with CMOS transistors have several good features. Channel resistance values are small and hence switching speeds are high, power consumption is low and immunity to fluctuations in the supply voltage lead to good immunity to noise. The chief disadvantage is that the packing density is smaller than ordinary MOS transistors.

Magnetic bubble domian memory

The development of the digital computer has been critically dependent on the continuing development of improved techniques for the storage of information. One of the most promising approaches in recent years is the magnetic bubble system. Offering accessible non-volatile memory in a very small package it presents many advantages. It is small, fast, consumes very little power and is becoming competitive with semiconductor devices.

The 'bubbles' are actually cylindrical magnetic domains whose polarisation is opposite to that of the thin film in which they are embedded (Figure 10.5). They are stable over a considerable range of conditions and can be moved from point to point at high speed. The medium in which the bubble moves is a thin layer of ferromagnetic material, magnetised perpendicular to the layer by an external magnetic field. Existence of the bubble domains is possible only in material that is magnetically anisotropic with the preferred (easy) direction of magnetisation perpendicular to the layer.

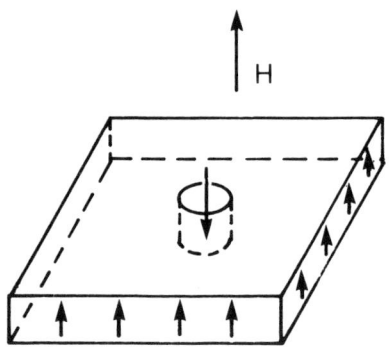

Figure 10.5. A bubble is a cylindrical magnetic domain of opposite magnetic polarity to that of its surroundings. It exists in a thin film of uniaxial anisotropic material under an external bias field

Fabrication

Production of a bubble device begins with a substrate (Figure 10.6) upon which an epitaxial film is grown and in which the magnetic domains are formed by the application of a d.c. magnetic field perpendicular to the film. The substrate is composed of a non-magnetic garnet in which the rare earth gadolinium and the rare metal gallium are incorporated. The composition of the bubble supporting epitaxial layer is a magnetic yttrium iron garnet. A typical film is 5 μm thick, the diameter of the bubbles is 3–5 μm, and the external applied field is about 4 kA m^{-1}. If the field is greater than this the bubbles disappear, if it is much smaller the bubbles change into an elongated filamentary domain and lose their efficacy. The external field is usually generated by coils. On the film are deposited patterns of Permalloy (a magnetic iron–nickel alloy) electrodes that define the path of the bubble domains in the presence of the external magnetic field.

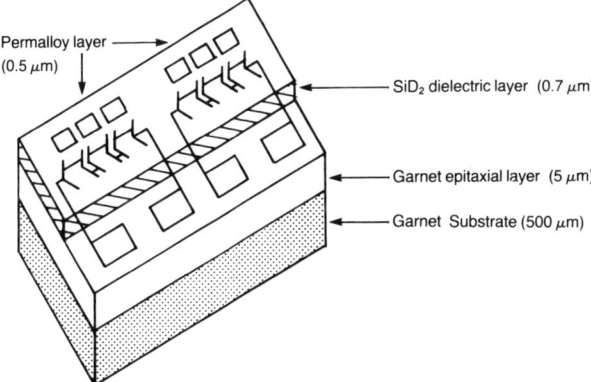

Figure 10.6. Schematic diagram of magnetic bubble memory chip (thickness of film in μm)

Arrays of the Permalloy are arranged in the shape of chevrons, T's and bars and serve the purpose of creating local variations in the magnetic field in order to localise and separate the individual bubbles. The entire stream flows at a regular rate in response to the periodic magnetisation of the Permalloy. An intermediate layer of SiO_2 between the epitaxial film and the Permalloy is inserted to improve the mobility of the bubbles. As regards movement, the bubble domains may be considered as magnets and hence subject to the influence of a magnetic field, the induced magnetism appearing as a moving train of poles which pull the attracted bubbles along the patterns of the Permalloy. The method used to produce movement is to induce magnetic charges thus producing a magnetic field gradient. In this aspect a magnetic field gradient is essential, for bubbles in a uniform field will not move but only contract or expand. The field gradients are produced by applying a TRANSVERSE uniform magnetic field to magnetise the Permalloy. Manipulation of the magnetic poles produced in the alloy allow controlled domain motion to be achieved. In effect the Permalloy patterns produce on the epitaxial layer a propagation track to define the bubble domain locations. Motion is controlled by a combination of the mutual attraction of the bubbles to the Permalloy and the time varying polarisation of the Permalloy elements by a rotating in-plane magnetic drive field (Figure 10.7). The propagation track is affected by a gradient of the magnetic bias field applied to the bubble.

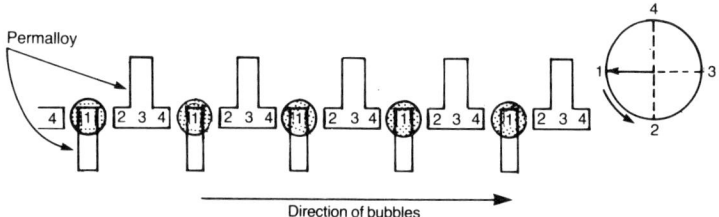

Figure 10.7. Bubble propagation. Anticlockwise rotation of the magnetic field causes a left-to-right movement of bubbles as they respond to the magnetic poles of the Permalloy pattern. With the field in position 1, the sites 1 form bubble attracting poles, with the field in position 2, sites 2 assume this function and so on

Information is carried by the stream of bubbles and voids, the presence of a bubble during, say, a 10 μs period constituting a logic '1' and, conversely, the absence of a bubble (a void) during the same period is a logic '0'.

The manner in which a computer bubble memory is usually arranged is illustrated in Figure 10.8. The information is stored in a series of small loops in which the bubbles move along tracks induced by the field gradients. These loops are connected to a large loop which is used for writing in, reading out, and erasing information. To enable the system to operate as a memory, provision has to be made for a bubble generator (G) a detector (D) and an annihilator (A). To enable bubbles to be switched from one track to another transfer gates are provided. Generators, transfer gates etc., are controlled by current pulses.

Data bits are introduced into and read from the major loop. This in effect is a unidirectional circular shift register that can transfer data to and from the top bit position of the minor loops.

Data blocks are accessed by rotation of the minor loops until the desired page of data is adjacent to the major loop, whereupon a page can be transferred from the minor loop to the major one. Capacity of magnetic bubble devices varies between 64K bits and 130K bits for

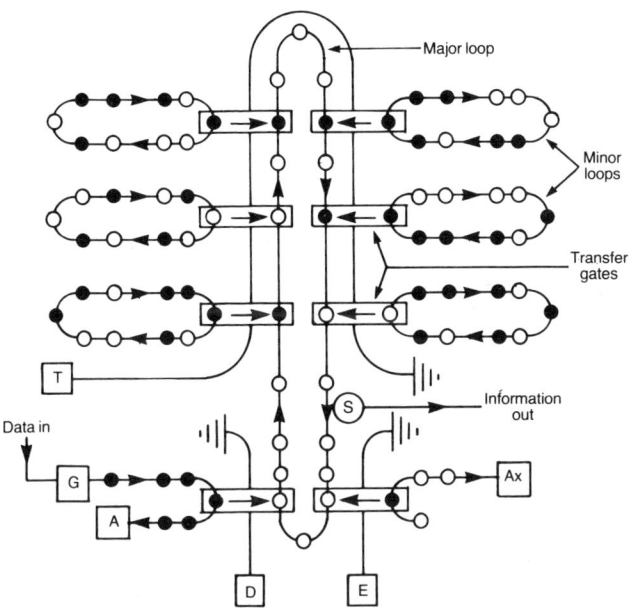

Figure 10.8. Layout of a magnetic bubble memory system. Information is stored in a number of minor loops in which the bubbles move along tracks similar to those shown in Figure 10.7. In the flowsheet, data (○) are being transferred from the minor loops into the major loop. New data are inserted into the memory by transferring selected bubbles from the bubble reservoir loop (G–A) in response to 'write' commands (D). Data are erased from the memory by operating E and transferring bubbles from the major loop into the annihilator (Ax)

a chip size of 5–8 mm². Storage density is between 2000 and 4000 bits mm^{-2}. Access times are in the region of 1 ms.

It should be noted that bubbles interact with one another like magnets when they get closer than about three diameters. These interactions tend to limit storage density. A variation of the bubble technique promises to overcome this limitation. Known as the bubble-lattice file or structureless bubble circuit, the bubbles are packed together in a hexagonal lattice, information being stored in the bubble walls. Distinction between binary states is determined by changes in the magnetisation within the walls of a single domain rather than by the presence or absence of a domain. The successful development of bubble lattice memories could theoretically multiply by ten the density of present memories.

It would seem that magnetic bubble technology appears destined for widespread use. It has however met with a cool reception in America and in fact at least three large US manufacturers have recently withdrawn from its production. It would seem that the circuitry needed for control is very complex — anywhere from 40-100 standard I.Cs being required to manipulate a 256k or a 1 megabit device. As a result profitability and bubble chip yields tend to be low and uneconomical. Japan however currently markets 64k bit and 256k bit bubble memory devices and has developed mass production procedures.

Software

Hardware is the term used to describe the mechanical and electrical component parts of a computer. The programs which control the internal operations and define in detail what has to be done are known as software. It provides the means for telling the computer explicitly what to do through a step-by-step sequence of instructions that form the program.

Machine language

The manipulation of data within the computer is performed in the binary code based on a scale of 2 in which the instructions are written in a sequence of 0s and 1s.

Because these sequences of 1's and 0's are acceptable by a computer as instructions for processing data and because they can be interpreteded by a computer they are referred to as machine language. Although a complete program can be written in this low-level language, the task is so tedious that an intermediate representation known as assembly language has been developed. This consists of groups of two, three or four letters known as mnemonics which relate directly to the machine code and describe the instructions. The purpose of the assembly language is to make it easier for the programmer to code programs and consists of a standard formulation of letters standing for a known word such as NOP for no operation, MPY for multiply or STE for store.

High-level language

The conversion of a program written in assembly language into machine code is done by the computer itself with a unit known as an 'assembler'. Assembly language programming is reasonably efficient when it comes to making the optimum use of memory space, but in order to make coding and programming still more flexible, 'high-level' languages have been developed. These languages enable programs to be written using an English/-mathematical vocabulary and are more readily acquired and used by programmers than the machine code.

The statement in such language usually corresponds to many statements in machine language. The process of translation is carried out by the computer under the control of what is known as a compiler. The compiler accepts as data the instructions written in the high-level language and convents them into the corresponding machine code instructions. The program that the programmer produces in the high-level language is called the source program and the translation produced by the compiler the object program (Figure 10.9) which is in machine language and after the correction

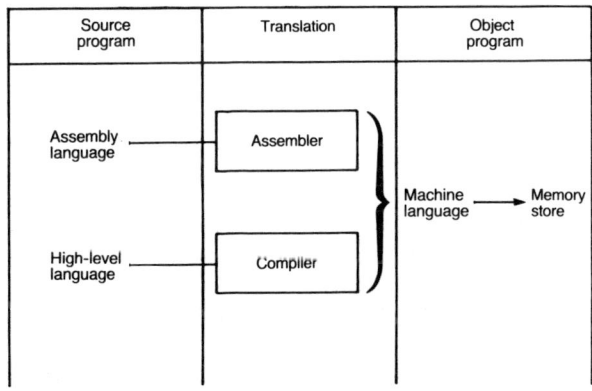

Figure 10.9. Software terminology relationships

of the inevitable mistakes (known as debugging) it is stored in the memory where it is held in readiness for use.

Both assemblers and compilers are language translators, the difference being that the compiler has to operate a more complex source language. A high-level language is directed towards the processing of a given class of data, for example payroll, stock control, insurance installments, and accordingly is said to be problem-orientated, whereas a low-level language is directed towards the requirements of the computer and hence is computer orientated. Because the high-level language is related to specific tasks rather than to a particular computer it is possible to create languages that are independent of the computer. Over the years there has been a proliferation of high-level languages for special purposes each with its own compiler for making it intelligible to particular models of computers. As a result 'cross software' systems have been developed to facilitate communications between computers, thus enabling a computer of one make to duplicate the action of another.

Amongst the high-level languages that are used for coding instructions the following are the most widely used.

Algol stands for *alge*braic-*o*rientated *l*anguage and is the accepted European program language and permits a programmer to write instructions for solving mathematical problems.

Fortran is the American equivalent and is the acronym for *for*mula *tran*slator. It is intended for scientific and engineering calculations and uses algebraic notation.

Thus the expression

$$X = \frac{b + \sqrt{(b^2 - 4ac)}}{2a}$$

would appear as follows in the Fortran notation

$$X = B + SQRT (B**2-4*A*C)/2*A$$

where/denotes division, the single asterisk denotes multiplication, the double asterisk exponentiation, and SQRT stands for square root.

Cobol (*c*ommon *b*usiness-*o*rientated *l*anguage) permits the writing of business data processing instructions. It is particularly suited to commercial application when alphanumeric data have to be stored and manipulated. It has found particular application in the field of librarianship and information sciences. PL/1 stands for programming language one and is an attempt to provide a universal language to cover both commercial and mathematical applications.

Coral was devised by the Royal Radar Establishment. It is a general purpose language based on Algol with some features from Fortran and is widely used in this country.

Input

Input information to the computer can be in a variety of forms — punched cards, punched paper tape, magnetic tape, magnetic ink (commonly found at the bottom of cheques) and optical scanning.

Punched cards

A common form of presenting data to a computer is by means of punched cards (Figure 10.10). The currently most widely used card measures 8.5 × 19 cm and is divided lengthwise into 80 columns and vertically into 12 rows. Of the 12 positions in each card, 9 are used to represent the numeric digits 1–9 and the other 3 known as zone positions are designated as 0, 10 and 11. Each column on the card can record one character, a digit, letter or symbol by means of a pattern of one, two or three holes unique to each character. Each card can contain a maximum of 80 characters.

The pattern of holes representing characters are known as a punching code which varies from one manufacturer to another. Numeric digits are represented by punching in on the appropriate column from 0 to 9, whereas alphabetic characters are represented by punching a zone digit (0, 10 or 11) in combination with a numeric digit.

A character is recorded by punching holes in the card by a machine that has a keyboard rather like that of a typewriter, depression of the keys on the keyboard causes holes to be punched column by column on to the card. The punched cards are then passed to the computer, the information they contain being transferred by means of a card reader. Here they pass under a photocell sensing device which is able to detect the holes in each card and generate appropriate signals in binary coding for transmission to the central processor and computer memory.

Punched paper tape works on the same principle of punching holes in a coded format but in this case consists of a long ribbon of paper 2.5 cm wide. Characters on the cards or tape can be read into the computer at speeds of 1200 characters per second.

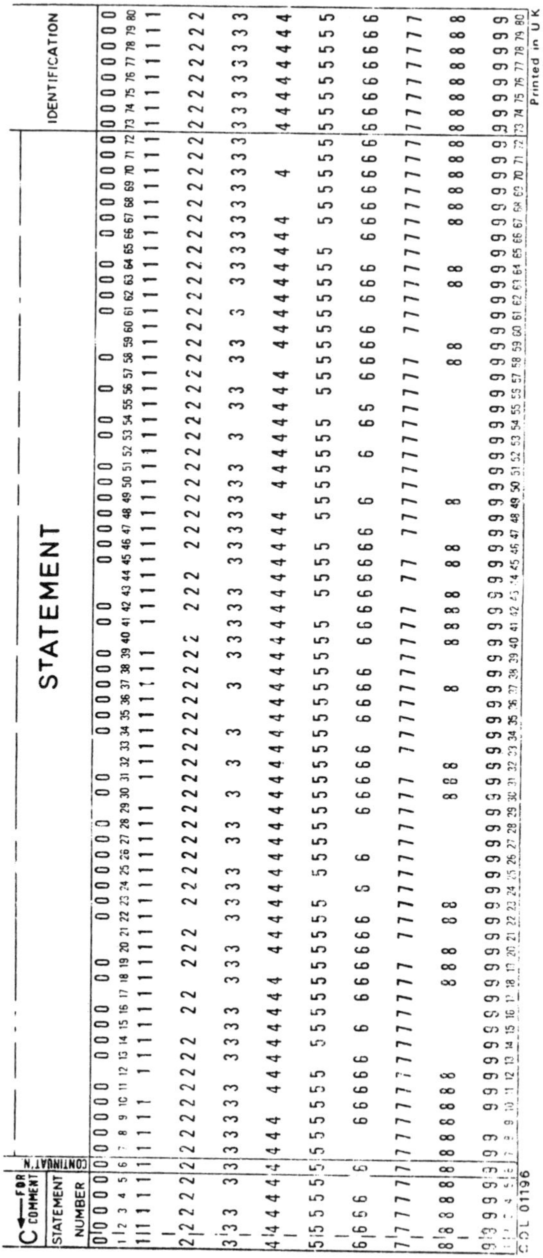

200 Digital computers

Figure 10.10. A punched card

After processing by the computer data are printed, generally on a large sheet of paper by a line printer. The writing performance averages 1500 or more lines per minute and with an average of 100 characters per line, the output rate is 2500 characters per second. This means that the entire contents of an average length textbook could be printed in about 3-5 minutes. When a printed copy is not required the output can be displayed on a visual display unit (VDU), a typewriter keyboard being located in front. This device is commonly used in remote terminals, a term used to describe input/output devices located at places distant from the computer itself. It is frequently seen, for example, in travel offices for the purpose of ascertaining whether a seat is available on a plane. In this case the operator at the remote end uses the keyboard to contact the computer, and within a few secnds the required information is displayed on the screen.

Computer arithmetic

In the binary system the arithmetic rules are comparatively simple:

Addition
$0+0=0$; $0+1=1$; $1+0=1$; $1+1=0$ with a carry over of 1
Multiplication
$0\times0=0$; $0\times1=0$; $1\times0=0$; $1\times1=1$
Division
$0\div1=0$; $1\div1=1$

Addition is performed in binary units following the same basic rules as those used in the decimal system. It should be noted that a 'carry over' occurs when a total exceeds one, whereas a carry over occurs in decimals after 9 is reached. For example the binary addition of 1010 (decimal 10) and 1011 (decimal 11) would be performed as follows:

Binary	*Decimal*
1010	10
1011	11
10101	21

In practice two numbers are added together by obtaining one number (bits) from the accumulator and the other number (bits) from the memory and adding these in the adder circuit. The result is placed in the accumulator, erasing the previous contents. Successive additions are carried out in this manner the accumulator assuming the new sums after each operation.

Subtraction is accomplished by adding the complement of the number to be subtracted to the minuend (upper number). This may be best explained by comparing it to the decimal subtraction using complements. The complement of a decimal number is the difference between it and ten. Thus the complement of 7 is $(10-7) = 3$. Seven subtracted from say 8 would be 1. The same result is obtained by ading to 8 the complement of seven and discarding whatever may be carried to the tens column. Thus the complement of 7 is 3 and $(8+3) = 11$ which on discarding the carried '1' leaves the same result as direct subtraction. In the binary system subtraction is carried out by inverting the number, that is changing 1 to 0 and 0 to 1 and adding the result. To subtract the binary number 100 (= 4) from 1011 (= 11); 100 is first inverted which yields 011, the complement is then formed by adding 1 to the invert, finally the complement 100 is added to the minuend 1011 yielding the answer (111) after discarding the carried '1'. Thus

$$\begin{array}{r} 011 \text{ (invert of 100)} \\ 1 \\ \hline 100 \\ 1011 \\ 100 \\ \hline (1)111 \ (=7) \end{array}$$

Multiplication. As shown in the table at the beginning of this section, binary multiplication consists of only four entries of which three are '0's since any number multiplied by 0 equals 0. Hence multiplication consists of copying the multiplicand where the multiplier is '1' and writing zeros where the multiplier is '0', shifting one place to the left each time.

For example 111 ($=7$) × 101 ($=5$)

$$\begin{array}{r} 111 \\ 101 \\ \hline 111 \\ 000 \\ 111 \\ \hline 100011 \end{array}$$ ($=35$ in decimal)

Multiplication can also be carried out by repeated addition: thus to multiply a number by five, add the number to itself four times.

Division is a combination of successive subtractions, the number by which the division is to be made (the divisor) being repeatably deducted from the number to be divided (dividend) until a difference less than the divisor is obtained. A numbering device in the ALU keeps count of the number of subtractions made.

As outlined above, by using the principle of complementing in binary arithmetic operations, it is possible to perform subtraction by addition. In this manner the four basic operations of arithmetic can be reduced to addition and hence the procedure is simplified because common circuitry can be used.

11 Microprocessors

The evolution of electronic technology over the past decade has been so rapid that it is sometimes called a revolution. The main driving force behind this revolution is the progress of large-scale integration in microelectronic circuitry in which smaller and smaller components perform increasingly complex electronic functions at ever increasing speeds and at ever lower cost. These features have made possible the transformation of a computer from a machine that must be built from many components into a component that can be incorporated into larger systems. This component, which constitutes what is known as a microprocessor, is the central processing unit (CPU) of a computer together with its associated circuitry scaled down so that it fits on a single silicon chip or chps holding many thousands of transistors. It is able to execute stored programs and is designed to interface with external devices. This is causing a tremendous shift to microprocessor-based instrumentation and new applications are being rapidly developed in industry, in banking, in power generation and distribution, in telecomunications and in scores of consumer products ranging from automobiles to electronic games. What makes the microprocessor important, however, is not the low cost or small size. Other IC's offer the same advantages; the truly revolutionary aspect is their versatility which stems from their programability. The function of microprocessors can be changed over again merely by modifying the software (reprogramming) of the same basic system. No longer do different circuits need different basic software. Today the same MPU chip may appear in a videogame, a laboratory instrument or in an automobile — programmed in each case to perform a different function in manufacturing, telecommunications or business. The economics are so compelling that MPU's have many applications not only where computing power was previously too costly but also where several MPU modules can now be coupled to monitor and control parts of existing industrial or commercial systems for which computer control was previously untenable.

General aspects

As the name implies an MPU is the CPU of a larger computer, shrunk down to a 'micro' size and in fact has been described as a 'computer on a chip'. It receives data in the form of strings of binary digits, stores the data for later processing to perform arithmetic and logic operations on the data in accordance with previously stored instructions and to deliver the results to the user through an output mechanism — such as an electric typewriter, a VDU or light-emitting diodes. The CPU can be viewed as a group of registers connected by a gating arrangement. In operation the CPU reads each instruction from memory and uses it to initiate various processing operations. It fetches instructions from memory, decides their binary contents and executes them. It can also rapidly obtain any data stored in a memory.

Built with ion-implanted n-channel silicon gate technology, the chip (about 0.5 cm × 0.5 cm) contains all the functions required for multi-instruction processing; an arithmetic and logic unit, instructions decode and address systems, instruction register, all of the clock and logic circuits required for timing and a full complement of data bits, input and output matrices and address bus drivers (Figures 11.1 and 11.2).

Signals on the input lines are the data input to the MPU. The data may come from switches, sensors, analogue-to-digital converters or keyboards. Inside the MPU resides the program which is a set of sequential instructions that determine how the input data are to be processed and what information is to be sent to the output lines as a consequence of the input data.

The output may be connected to actuators, digital displays, digital-to-analogue converters etc.. The lines used to carry data to

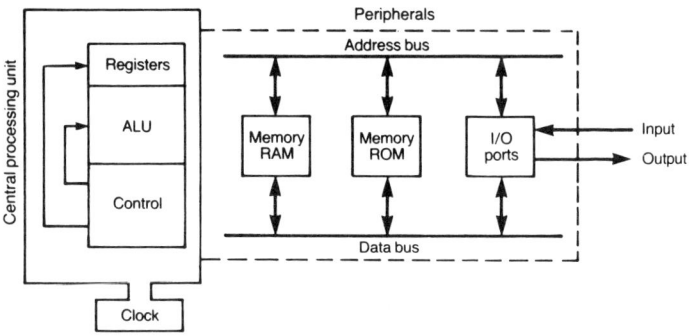

Figure 11.1. Illustration of the salient hardware features of a single-chip microprocessor. The three distinct parts of this type of architecture are:

(1) ALU which carries out such simple functions as addition, subtraction and logic AND, OR etc., on data received over the bus.
(2) Control section which organises the actual sequence of events in the CPU, the clock generating the timing information for the whole processor system.
(3) Registers are small storage locations that, upon command, extract from a memory bank, perform the programmed operations and then return the data to the memory or output device. When combined with memory chips for storing operating instructions and for controlling the input/output functions the unit becomes a microcomputer. This is indicated in the diagram by the area enclosed by the broken lines

and from the MPU are known as buses (Figure 11.3) and are defined by data path width or word size, 4, 8 bit etc.. The address bus is unidirectional and used to transmit an address from the MPU to the memory, input or output unit. The data bus is bidirectional and is the path for data flow. In addition to the data and address buses, MPUs also have a set of control lines, both input and output, over which signals travel to maintain timing and synchronise the operation of the MPU to the operation of the external circuitry.

Although the MPU in the sense of a small processor for digital computing equipment has been known and used for many years, the era of modern monolithic large-scale integrated (LSI) MPUs, in contrast to the earlier small-scale and medium-scale integrated machines, is a direct result of the extensive advances in IC design in microelectronic device processing technology. The perfection of n-channel MOS LSI technology has made possible the processing of circuits having higher component densitites, lower power

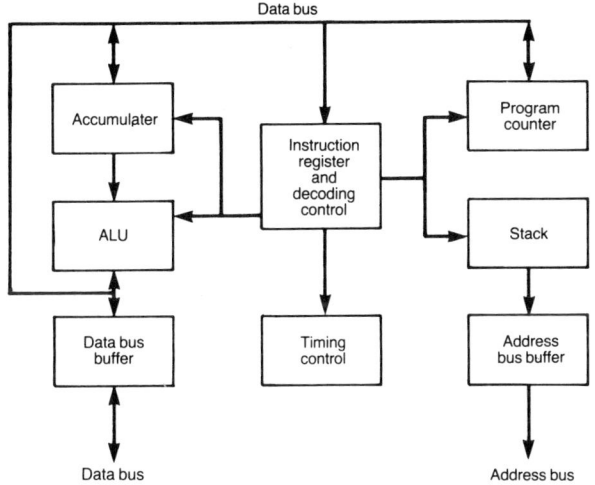

Figure 11.2. Block diagram of a CPU illustrating in more detail the architecture. The accumulator works in close association with the ALU and can be regarded as the main working register into which data and results are written and processed, and from which they are subsequently dispatched to memory or output ports. The instructions which form a program are stored in the Program Counter enabling instructions to be addressed and brought down from memory in the proper sequence. The Stack is a sequence of registers used to store the program counter's contents and is sometimes referred to as a LIFO, or last in first out. Data Buffer is a storage device that compensates for differences in rate of data flow when transmitting information from one computer to another

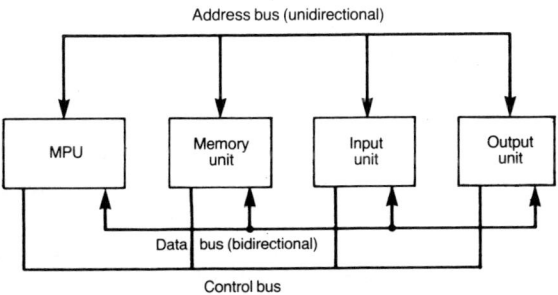

Figure 11.3. Bus structure of a typical MPU

requirements and better production yields as against the equivalent bipolar circuits.

These advances led to the introduction in 1971 of the 4-bit MPU with about 500 gates on a single silicon chip. This MPU when combined with a memory control, a temporary memory store and a master clock constituted an entire micro-computer that originally could sell for less than £25. During recent years many different large-scale MPU single-and multi-chip arrangements have been introduced with word-lengths ranging from 9 to 16 bits and 6000 gates. Looking further ahead the technology will provide for 10 000–20 000 gates which will allow a 32-bit processor with 4096 bytes (4K) of RAM storage on a single chip, giving it the same word length as a large mainframe computer.

A microprocessor is not a complete computer. It has only real value when it is used as a component part of a computing system. To make a computer out of a MPU requires the addition of memory for its control program, plus input and output circuits to operate peripheral equipment. In effect it is the control and processing unit of a small computer (microcomputer).

In addition, various other units have to be incorporated. Power supply is needed to deliver supply voltages to both the MPU and its associated circuitry. Clock circuitry is required to generate proper timing signals for synchronising the system's operation. This usually consists of an oscillator composed of a quartz crystal plus two capacitors which set the operating frequency.

Memory

Although it is possible that the necessary memory may be contained on the MPU chip itself it is much more usual for it to be external to the MPU. Two types of memory are used, read-only memory (ROM) and random-access memory (RAM). ROMs are usually pre-programmed during manufacture, often according to the customer's specifications. The data stored in ROM are vital to the operation of the microcomputer and must not be lost when the supply is turned off. The data are therefore programmed in the form of firmware. PROMs (user programmable ROM) provide the same functions but internal bit patterns can be set by the user. However, the material still cannot be modified once it is written. A type known as EPROM (erasable programmable read-only memory) gets round this problem. The program may be erased by ultravoilet light and the EPROM reprogrammed. Random-access memory allows for information to be written and modified as well as read. An important change that MPUs have brought to the design function is a substantial shift in the delineation. The older technique of implementing the logic of a product in the interconnection of standard logic gates has been in part replaced by interconnection with the logic stored in a ROM. This has permitted the designer to place nearly all the logic in a few integrated circuits rather than diffused throughout the system. With the logic concentrated in only a few components a high degree of flexibility is possible.

Circuit elements

The majority of MPUs use MOS transistors rather than bipolar transistors. The main advantage is their higher packing density, allowing more functions to be placed on a silicon chip of a given size than could be attained using bipolar circuitry.

The p-channel MOS was the first device used for the fabrication of monolithic semiconductor microprocessors and was the result of use in calculator projects. The advantages were simplicity, proven track record, good yield and availability. The main disadvantage was the limitation in speed because of the inherently lower carrier mobility of the holes. The speed disadvantage is overcome by the use of n-channel MOS devices, since Electrons move faster than holes. Although the earlier n-channel microprocessors used enhancement mode transistors, the modern ones use depletion mode which has eliminated the need for an extra power-supply voltage.

More recently complementary MOS (CMOS) microprocessors have appeared and offer the advantages of high performance, extremely low power dissipation, high noise immunity and wide operating range with respect to supply voltage.

Bipolar devices

Bipolar technologies have been used and offer the advantage of very high speeds. Transistor–transistor logic (TTL) devices with Schottky clamping to improve packing density and reduction of power per gate can be used as auxiliary components in a microcomputer system for the control of memory and other functions. Nevertheless a substantial gap remains between bipolar and MOS devices. One candidate to fill this gap would appear to be integrated-injection logic (I^2L) which overcomes the traditional drawbacks of bipolar devices, namely low packing density and high power dissipation per gate and in fact can compete with CMOS at the same speeds (5 ns per gate). I^2L seems potentially to have overtaken other technologies to a significant extent. In addition, it has the capability for direct interface with analogue signals which is of great importance in, among others, the automotive and consumer markets.

Software

Even the most elementary computer needs a set of instructions — which is what software amounts to. producing these instructions is a highly skilled labour-intensive job. These instructions are stored as a sequence of logic '1's and '0's in memory and make up the machine language. Programs of 100 words or more create problems in binary bit patterns and hence the advent of assembly and high-level languages which are human-orientated varieties of machine languages and are more easily understood by people than machine instructions. These languages have of course to be translated into machine language before they can be executed. Assembly language is the most common language used in writing programs for microprocessors. Use of high-level languages can in most cases reduce program development time and both compilers and interpreters are available as high-level language translators. The most common interpretive language is BASIC (Beginners All Purpose Symbolic Instruction Code) whilst the most common compiler languages are Algol, PL/M, Fortran, Cobol and Pascal.

As microprocessors have become more sophisticated the cost of producing the software for a given application has risen rapidly. As a result the hardware costs of computers are now overwhelmed by the software costs. Software development costs about £5 per written line of program or several times the cost of an IC chip including design, coding, testing, integrating and documentation. Some 70% of the total cost of a software product occurs after it is delivered and goes into debugging (trouble shooting) and maintenance. Software maintenance alone takes 50-80% of the data-processing budget at a typical installation. All in all, estimates suggest that the worldwide capacity to write programs is growing by 18% a year, which is several percentage points slower than the rate at which MPUs are being installed.

Many kinds of remedies are being tried to redress the situation. One is to increase the use of high-level languages which enable programmers to write software faster. Another ploy called structured programming breaks down programming tasks so that they can be shared more efficiently between programmers. A third approach is to wire standard programmes into the hadware of computers and thereby, in theory, mass-produce the software. None of these remedies, however, overcomes the basic problem; programming work is a creative activity not vulnerable to measurement, and therefore to control, and to a large extent has to be individually tailored to the needs of each customer.

It would seem that these difficulties could put a brake on the spread of computing power in the next few years. For instance, delays in developing software have forced one large company to postpone a computer system for small businesses which was a key part of a new network of distributing computing. Another large company has been trying to hire 2500 experienced operators — who are becoming increasingly expensive — and has been forced to spread its recruitment world wide.

Assembly and high level languages

Although most microprocessors today use low-level assembly language many manufacturers are nevertheless adapting high-level languages for microprocessors, since they take less time to write in machine code and they do not have to be relearned for each new processor. Of particular interest in this respect is the Pascal, introduced in 1971 by a Swiss computer scientist. It was first used on mainframes but has recently spread to MPUs. Its advantages are that it defines its terms more precisely than conventional BASIC, Fortran and Cobol and the software is therefore easier to understand, debug and modify. A Pascal intermediate language or

P-code can now be written in standard form for any MPU and then interpreted separately for each processor machine language. The strategy reportedly can speed up markedly the software development time.

Applications

As microprocessors diffuse into a myriad of applications in factories, offices and homes it is difficult to visualise any aspect of contemporary life that has escaped its impact. The flood of microprocessors reaching the market combined with the rapid rate of innovation shows that any attempt to catalogue their usage in a text of this length is impossible and hence only a few areas can be illustrated.

Creative people in the entertainment and consumer fields have been among the first to recognise their potential and have flooded the market with MPU-based pinball games, TV games, microwave ovens, sewing machines, cash registers, washing machines etc.

Hand-held microprocessor products are becoming more sophisticated. programmable calculators are resembling the first generation of personal computers which can handle not only numbers but also converse in words and alphanumeric symbols. Language translators are available which speak a multiple combination of words in different languages. Plug-in ROM modules are available for English, Spanish, French and German. Peripheral entertainment equipment for the TV receiver which gives the users access to a wide variety of video games has also become available.

Land, air and sea transportation systems have been invaded and improved by mircoprocessors. In the automobile industry, car manufacturers are including microprocessors to measure temperature, engine speed, pressure and the oxygen content of the exhaust gases — relative either to fuel economy or pollution control, or both — and to readjust control of the engine accordingly. Spark timing and choke control are also being taken care of by the MPU.

Modern jet aircraft depend on a variety of systems for navigation, communication, passenger comfort, safety and engine control. MPUs monitor these various systems and transmit their data to a central computer which generates the control signals needed to keep the systems functioning correctly. This leads to a great saving in cable-costs, increased reliability and lower maintenance costs.

Microprocessors in industry are being used as controllers. Typically, several MPU's each handling a specific function are linked together by a minicomputer, which performs the larger data-manipulation task. A large volume aplication like the cutting and shaping of metals illustrates this usage. Multi-purpose turning and machining centres capable of changing tools automatically have now appeared on the market.

New robots for parts assembly are proving attractive for industrial batch assembly. Until recently most robots were used for such operations as welding, parts transfer, forging and the operation of punch presses. Recent robots have included versions which can recognise poor metal welds and improperly positioned machine tools as well as do the work. In short, spearheaded by MPUs electronics is controlling and monitoring processes far more efficiently and quietly than electromechanical relays — and with minimum human supervision.

Future trends

Today, microprocessor trends are in the direction of computers complete with memory and input circuits on a single chip, of more powerful processors, simpler power supplies and higher speeds. The advent of the 32-bit MPU will mean that by the end of this decade the microprocessor will be moving into its fourth generation. The progression of generations has been 4 bits, 8 bits, 16 bits and now 32 bits (the number of bits referes to the size of word which the MPU can undertake at any one instance). The 32-bit chip will give access to a vast amount of data and will contain more than 20 000 gates and provide the computing power of a large mainframe computer at a fraction of the cost.

The application of very large scale Integration (VLSI) to MPUs is now taking shape. One VLSI project visualises much more complex processors of 100 000 gates on a single chip, interconnected to similar processors on other chips which would enable the ensemble to exceed the power of present large mainframe computers or even to replace them. In addition new processors could be added to expand the processing power without hardware or software redesign.

Microsensors

Transducers that are fabricated by silicon IC technology and can fit on the same chip with a MPU are now being developed and will allow MPUs to be used in certain specialised high-volume consumer applications. These include automobiles, refrigerators, air conditioners, cameras. Micriprocessors have scarecely touched these markets because they hitherto have lacked the inexpensive reliable sensors that can be fabricated by silicon IC technology to measure such variables as pressure, temperature, humidity, force, acceleration and gas flow.

A sensor operating on the principle of the photovoltaic junction effect is shown in Figure 11.4. In this, light falling on a photoconductive sensor causes a photovoltage to be produced. The detector operates with a reverse-biased voltage applied to the diode,

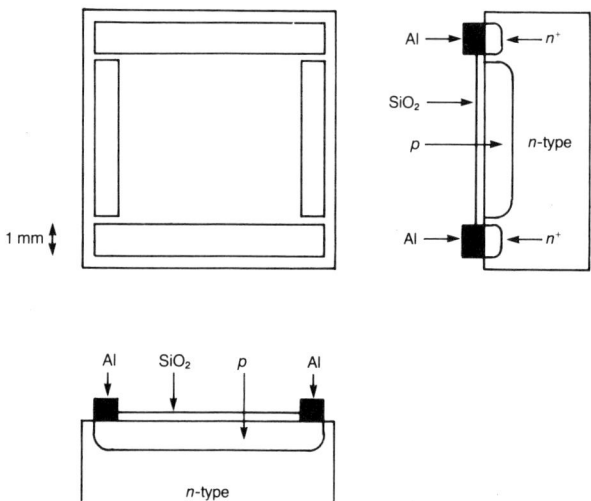

Figure 11.4. Plan and cross section of a photovoltaic junction sensor which can fit on the same chip with a microprocessor (Spectrum, February 1980, p. 44)

to prevent forward biasing when the diode is illuminated. The light sensititve element of the detector is made by diffusing a layer of borondoped P-type silicon into a substrate of phosphor-doped n-type silicon. The element is framed by 1 mm wide aluminium contact strips. As regards usage the detector may be applied in adjusting laser beam playback systems for video disc players; in alignment systems for photolithography; in pattern recognition; and in target seeking systems for air-to-air or surface-to-surface missiles.

It is evident that the integration of sensors and signal processing on the same chip will generate a new class of microprocessors.

12 Very-large-scale integration

The wide range of modern digital electronics — from computers to communication satellites — has only become a practical possibility because of IC technology and its extension to large-scale integration (LSI). Significant progress has been made in this field in the past few years; a single silicon chip can now accommodate an entire microprocessor. Advancement in semiconductor-chip density and complexity beyond LSI has, however, received a major boost in

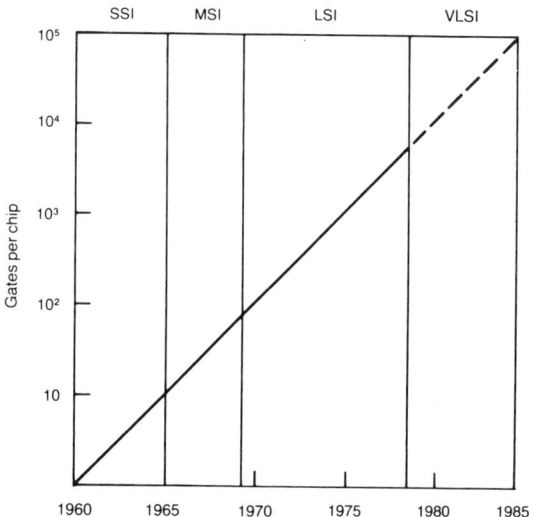

Figure 12.1. The evolution of ICs in 20 years based on the increase in the number of gates per chip

recent years. Fabrication techniques — ever the key to future integrated circuits — have advanced to the point where achievement of submicrometre sizes has now been realised. This accomplishment has ushered in a new semiconductor era, characterised as that of very-large-scale integration (VLSI), with the possibility of incorporating more than 100 000 gates per chip. The main driving force behind this innovation is the rapid growth in the use of microprocessors and the need for information storage (memory).

The era of integrated circuits began in the late 1950s with the discovery and development of the planar silicon technique which opened up the way for the integration on the same semi-conductor chip of an ever-increasing number of circuit functions. In twenty years IC technology has advanced from SSI (small-scale integration, 10 gates per chip) via MSI (medium-scale integration, 100 gates per chip or 256 bits of memory) to LSI (large-scale integration, 1000 gates per chip) and now to VLSI (very-large-scale integration, 10^5 gates per chip (Figure 12.1).

Cost reduction

One of the most important benefits of IC technology is its relentless trend towards lower and lower costs. Historically, costs have decreased aout 30% every time cumulative volumes have doubled. The cost of a given electronic function has been declining even more rapidly than the cost of ICs, since the complexity of the circuits has been increasing as their price has decreased. For example, the cost per bit (binary digit) of random-access memory has declined an average of 35% per year since 1970 when the major growth in the adoption of semiconductor memory elements got under way. (Figure 12.2). These cost declines were accomplished by the integration of more bits into each IC. The primary means of cost reduction has been the development of increasingly complex circuits that lower the cost per function. One of the main technical barriers to achieving more functions per circuit has been production yield. More complex circuits result in larger devices and a growing probability of defects so that a higher precentage of devices have to be scrapped. Technology development have concentrated primarily on increasing the production yield by increasing reliabilty and by

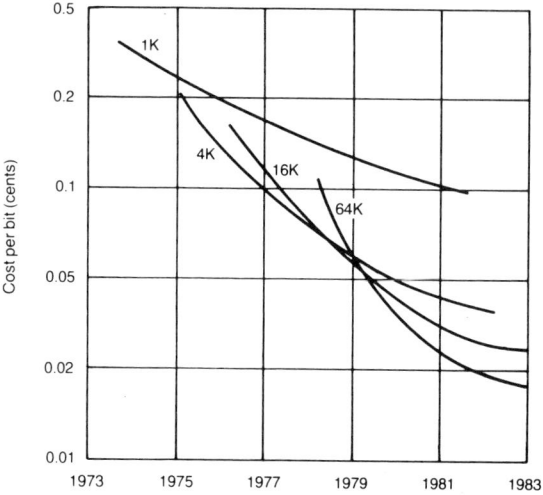

Figure 12.2. Memory component cost trends. Cost per bit of computer memories for successive generations of RAM circuits capable of handling from 1024 (1K) to 65 536 (65K) bits of memory (IEEE Transactions, April 1979, p. 289)

reducing dimensions. Meticulous attention to process control and cleanliness have been necessary to reduce defect density. A dust particle in any critical process is enough to make a device worthless. Special clothing is worn to protect the environment from dust and the air is continuously filtered to keep the dust level at a minimum.

Size reduction

The second most important trend is the continual shrinkage of IC dimensions (linewidths and spacings). Dimensions have historically shrunk by a factor of two every five years. This reduction in size has come not only from ingenuity in designing the circuits but also from the continuing miniaturisation of the circuit elements and their interconnections. Reduction of the dimensions of the base circuit elements which enable an increase in the circuit within a given area has been accomplished by improving the resolution of the photoengraving process. Optical limits have now been reached as dimensions in the circuit pattern enter the range of only a few wavelengths of light and methods in which electron beams and X-rays are substituted for visible light have been developed in order to reduce the dimensions even further. It would appear that this processing by the new lithographic techniques will make possible the fabrication of chips containing 10^7 individual transistors. One such chip will contain more functions than today's largest computer. Reduction in size of the circuit elements not only reduces the cost, but also improves the basic performance of the device. Delay times are directly proportional to the dimensons of the circuit so that it becomes faster as it becomes smaller. Similarly the power is reduced within the area of the circuits.

VLSI requirements

The principal requirements of a digital IC technology to make possible the development of ultra-high-speed, VLSI circuits are:

(1) Very high density (low chip area per gate).
(2) Low gate power dissipation.
(3) Low dynamic switching energy (speed–power product).
(4) High speed (very low gate-progagation delay).
(5) High process yield.

The origins of most of these requirements are obvious. Clearly, large numbers of gates (10^4 to 10^5) cannot be placed on a reasonable size (~ 1 cm^2) chip unless the gate areas are small (<1000 μm^2 per gate). The power per gate must be low (<< 1 mW) if chip dissipations are to remain manageable.

These requirements can be met by some form of integrated-injection logic (I^{2}L), n-channel MOS, CMOS or gallium arsenide. As Table 12.1 shows, the technologies available are not lacking. MOS technologies show much promise for VLSI for many reasons. The number of process steps in the fabrication of a MOS transistor is much smaller than for the other types, in fact it was this inherent simplicity that provided the impetus to work on MOS transistors in the first instance. MOS devices need little power and are self-isolating, thus charge can be stored on their capacitive gates, making dynamic logic and memories possible. MOS with its low dissipation,

Table 12.1 Technologies available for VLSI

Property	TTL	Bipolar ECL	I²L	pMOS	MOS nMOS	CMOS	Gallium arsenide
Number processing steps	18–22	19–23	13–17	8–14	9–15	14–17	16
Number of components per two-input gate	12	8	3–4	3	3	4	2
Packing density (gates/mm²)	10–20	15–20	75–150	75–150	100–200	40–90	300–1000
Propagation delay (ns)	10	2	20	100	15	20	0.07
Speed–power product (pJ)	30–150	15–80	0.2–2.0	50–500	5–50	2–40	0.01–0.1

Source: Lockhead Microelectronics Centre

high switching speed and noise insensitivity would seem to be a firm candidate for VLSI.

As regards bipolar transistors, these demand more processing steps and tend to dissipate more power than other technologies. I²L, however, with its low dissipation (0.05 mW), high switching speeds (1 pJ) and high packing density, offers many attractive features for VLSI and in fact of all bipolar technologies I²L has the best chance to meet the density, performance and reliability requirements for VLSI.

The 64K-bit RAM

The main driving force behind VLSI is the rapid growth of the microprocessor and information storage (memory) and in fact the technology of digital storage is the most rapidly developing area of VLSI. In fact it may be stated that the era of VLSI was heralded in 1979 with the announcement of the 64K-bit dynamic RAM.

RAM's have so far been the proving ground for VLSI and minimum geometries now in production are 2–3 μm typified by the 64K-bit dynamic RAM and the 16K-bit static RAM (Table 12.2). Channel lengths of 3–5 μm are common in the high-speed logic circuits of microprocessors. Projected for the (mid) 1980's are minimum geometries of 1 μm on 1-cm² chips. The packing density of RAMs increases as the inverse square of the smallest length and so the minimum geometry of the 1-μm chip corresponds to a memory with four to nine times the density of the 64K-bit RAM.

Table 12.2

Memory type	Process	Typical access time (ns)	Power (mW)
16K dynamic RAM	nMOS	80–150	200
16K static RAM	nMOS	45	550
64K static RAM	I²L	90	300
64K dynamic RAM	nMOS	125	250

A dynamic RAM is a volatile memory and stores data only as long as power is supplied. A clock must continuously refresh stored charge for a dynamic memory to function properly. Static memory on the other hand is non-volatile and does not require clock refresh circuitry but is slower than dynamic and requires more chip area. RAMs depend for their operation on the ability of the gate capacitance of MOS devices to retain their charge for relatively long periods (long that is, compared with the time for a complete cycle of operation within the system which may be only a few hundred μs in a computer system).

The 1k-bit dynamic RAM was the first to achieve full production status in the early years of the past decade. Since then 4K-bit and 16K-bit RAMs have reached volume production. The evolution has been made possible by advances in circuit design, process engineering and photolithography. The 1K-bit RAM employing a p-channel fabrication process used a three-transistor memory cell with an area of approximately 4000 μm^2. The 4K-bit RAM used a one-transistor memory cell and n-channel technology. This reduced the size of the memory cell to about 1000 μm^2. For the 16K-bit generation a second polysilicon layer was added to give a more efficient layout for the memory cells reducing the area to 450 μm^2. It should be noted that throughout this evolution the overall chip size of each generation has remained almost constant by reduction of the minimum feature size. This trend points the way for the 64K-bit RAM, using n-channel silicon gate technology, with an area of 200 μm^2.

By the early 1980s second generation 64K-bit RAMs will have appeared followed closely by 256K-bit RAMs. A bubble memory, 256K-bit device is now in production. Speed-power product per bit will be reduced by a factor of 20 (relative to the 64K-bit RAM). Feature sizes of 1 μm with isolation between active elements of 1–1.5 μm will be used, made possible by direct-slice writing techniques using electron-beam lithography and advanced dry processing. It is possible that the 64K-bit RAM will be the last generation of memory components to use the Silicon transistor cell concept — or for that matter the MOS transistor as we know it today.

Manufacture of the 64K RAM chip has presented difficulties. Production yields are believed to be below 5%. A technique recently introduced known as 'redundancy' has proved promising in increasing yields. In this method a superfluity of cells are incorporated on the chip. In the course of testing the faulty cells are deactivated by triggering fuses. The circuit is then rerouted to one of the extra fault free cells.

At present 64K RAM's are available only in limited quantities and consequently cost very much more than 16K RAM's.

Role of silicon

The primary material of the IC is silicon and the development of ICs has depended on the invention of techniques for making the various functional units on or in a crystal of this semiconductor material. The dominant role of silicon as the material for IC is attributable to its unique combination of properties. It is a semiconductor with excellent and easily regulated electrical properties; it is possible to make perfect single crystals of silicon — by the Czochralski method of crystal pulling from the melt — with a diameter of 10 cm and a content of undesirable impurities between 10 and 0.001 p.p.m. Finally, and most importantly, it can be oxidised to form a thin, dense and homogeneous layer of oxide that is very strongly bonded to the silicon surface. This oxide plays a major role both in the fabrication of silicon devices and in their operation and is particularly important in the fabrication of ICs because it serves as a mask for the selective introduction of dopants.

The SiO_2 layer is grown at temperatures in the range 1000 to 1200 °C, in an atmosphere of oxygen or water vapour, the required thickness being accurately controlled by selecting the appropriate time and temperature of oxidation. For example, a layer of oxide 0.1 μm thick will grow in one hour at a temperature of 1050 °C. An important aspect of the oxidation process is the low cost. Several hundred silicon wafers can be oxidised simultaneously in a single operation. A small process-control computer monitors the temperature, directs the insertion and withdrawal of the wafer and controls the internal environment of the furnace. Application of high-pressure oxidation has recently been introduced for achieving an accelerated oxide growth and preparing thermal oxide at reduced temperature.

Alternatives to silicon

In spite of the outstanding properties of silicon the high speeds of supercomputers are approaching the theoretical limits of silicon and even of the speed of light itself. The problems and limitations of silicon electronics have led to a search for alternative semiconductors and other techniques. Of these gallium arsenide cryoelectronic devices and bubble device structures are perhaps the most promising.

Gallium arsenide

Recent advances in the state of GaAs IC fabrication techniques have made possible the demonstration of ultra-high performances. GaAs digital ICs with gate areas and power dissipations sufficiently low to make VLSI circuits available. GaAs offers major performance advantages over silicon IC fabrication. The high mobility (GaAs 8600 electrons cm^2 V-s^{-1}, Si 1350 electrons cm$^{2\ \text{V-s}^{-1}}$) and electron velocities in short channel GaAs MESFETs (Schottky-gate, MESFET being an acronym for metal semiconductor FET) with trans-conductances or drain current about six times those of silicon n-channel FETs at equivalent gate biases (Figure 12.3). This FET

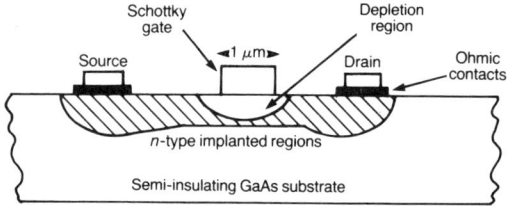

Figure 12.3. Cross section of a planar GaAs Schottky gate FET. It comprises a thin (10 μm) n-type action region joining two ohmic contacts with a narrow metal Schottky barrier gate separating source and gate. Operation is similar to that of a normal JFET with the conducting n-channel confined between the gate depletion region in the semi-insulating GaAs substrate

performance aided by the availability of a semi-insulating GaAs substrate makes possible much higher switching speeds for GaAs ICs in comparison with silicon ICs. Characteristics of the GaAs IC demonstrate that it is in the range required for ultra-high-speed VLSI, being one to two orders of magnitude faster and consuming well over two orders of magnitude less power than the current Si polar implementation. This achievement of very-high-speed GaAs

VLSI circuits allows new dimensions in on-chip computer power. A 10^5 gate GaAs chip with speeds of r_d 100 ps operating in some type of pipeline processor would have, assuming a fc = 1/5rd clock frequency, a gate count-clock rate product of $2 \times 10^{14-1}$ — a truly prodigous computation rate.

Josephson tunnel junction

The Josephson junction discovered by a British physicist (B. Josephson) is a switching device that harnesses the transistor between a super-conducting tunnelling state and a normal tunnelling state in response to a small change in a magnetic field.

If two superconducting materials such as niobium–zirconium or niobium–titanium form a junction with a very thin oxide layer (4 nm thick) sandwiched in between, the oxide, normally an insulating barrier, allows a current to flow with no appreciable voltage drop by a tunnelling mechanism. Tunnelling refers to the ability of electrons under certain conditions to penetrate energy barriers they would ordinarily lack the energy to surmount.

To achieve superconductivity (zero resistance) the device must be cooled to 4 K (–269 °C) by immersion in liquid helium. The device offers a dramatic potential to the introduction of a new phase of computer technology. Operating at temperatures close to absolute zero and at the very low voltages characteristic of the superconductivity state it is capable of switching speeds many times faster than present semiconductor circuits with a dissipation of less than one-thousandth the power of existing LSI devices. The reduction in power consumption makes it possible to produce very dense circuits without creating problems arising from the necessity to dissipate heat. The fabrication of the junction tunnel insulating barrier less than one-millionth of a centimetre thick has presented some difficulty, but a variety of Josephson logic circuits have already been built and tested in the laboratory. Cooling by immersion in helium is entirely feasible for helium cryostats are already in use for a variety of applications.

Magnetic bubbles

Magnetic bubble, memories already considered on page 191, have also been introduced to VLSI. Although slower than semiconductor memories they have vastly greater capacity and are non-volatile, retaining this data in the event of power failure. From the time they were discovered their application as memory elements seemed manifest, the presence of a bubble representing a logic 1, its absence a logic 0. The theoretical densities possible — at a time when semiconductor memories were packing but a few hundred bits on a chip — were almost incredible, hundreds of thousands of bits per square centimetre.

One megabit and 256K memories are now being produced by what is known as field-access technology. In this method the bubble-containing layer is covered with a thin-film pattern of Permalloy (NiFe) islands lying in the plane of a rotating magnetic field, generated by two coils. The oscillating changes of magnetic polarity attract the bubbles, which lie under the Permalloy islands, and move them from one island to another. What is actually propagated is the domain wall — not a single unchanging 'bubble'. The field-access propagation method is limited to bit densities of 1 megabit. Large capacity up to 4 megabit are possible by a method known as the bubble-lattice file or structureless bubble circuit.

In the field-access method 1s and 0s are stored as the presence or absence of bubbles. The bubble lattice circuit stores information in the form of wall states present on a single bubble. These states, infinite in number, are caused by twisting the magnetisation along the cylindrical wall of the bubbles. The distinction between binary states is thus determined by changes in the magnetisation within the wall of a single domain rather than by the presence or absence of a domain. Thus one bubble can store both a 1 and a 0 and hence the bubbles can be packed very tightly together, giving an increased density.

Obstacles to VLSI

(1) Manufacture

The challenge here is to etch smaller and smaller circuit lines on to the chip, every ten-fold decrease in the width of those lines leading to a hundred-fold increase in the number of circuits that can be put on the chip. At present the line widths average 2 μm, and it would appear the physical limit of miniaturisation is 0.3 μm wide. A number of promising techniques for achieving very fine resolution are at present being investigated and lines down to 1 μm wide have been achieved.

(2) Packaging

As chips are stacked with more and more circuits, more interconnection (wiring) is required to connect them to each other and to the outside world. These inter-communications use up more and more of the chip area and in fact the area occupied by the on-chip wiring is rapidly approaching that occupied by the circuitry. The electrical resistance of the interconnecting paths rise as geometries are scaled down because the cross section of the wiring is reduced which leads to higher power dissipation and lower speed.

(3) Thermal barrier

In use each circuit requires the generation of a very small electric current. The greater the density of circuits the greater the density of the power generated until chips begin to generate so much heat that they seriously impair their own performance. Thermal barriers thus come into play due to limitations in the ability to conduct heat away.

(4) Testing

Finally, the inspection and testing of chips holding over 100 000 circuits in order to discover the inevitable mistakes is an awesome task which is becoming more apparent to the industry. Automated computer test equipment is having to be programmed to deal with this new order of complexity.

Summing up it would appear that with the introduction of VLSI the electronics industry is achieving an increasing capability to integrate circuitry. Economies of scale are expected in storage technology leading to the use of a variety of technologies and to cost advantages. The revolution in such products as pocket calculators and wristwatches was brought about by advances in LSI technology and it is not difficult to speculate that VLSI devices will also have an important impact on consumer electronics in the future.

13 Optoelectronic components

Optoelectronics is concerned with the conversion of optical radiation into electrical energy (photoelectronics) or the emission of optical radiation upon the application of current or voltage (electroluminescence).

Photoelectronic devices

The photoelectronic effect is concerned with the change in electrical characteristics caused by radiant energy in the form of infrared, visible light and ultraviolet. Of those materials concerned with the technological exploitation of this relationship between optical and electronic phenomena, the semiconductors are an important class of photoactivated and photogenerative materials and have industrial application in machinery control, information transmission, digital processing, power generation etc.

Their usefulness as photoelectric devices arises from the fact that semiconductors are characterised by a narrow gap between the filled valancy band and the conductor band. Their ability to conduct electricity depends on electrons being excited into the conductor bands by various forms of energy. Light is one form of energy capable of liberating the current carriers and hence the absorption of light by a semiconductor frees the charge and gives rise to electrical current.

The effects whereby irradiation in the form of light produces variations in the electrical properties of the materials irradiated can be classified under three principal categories, namely photovoltaic, photoemission and photoconductivity effects.

Photovoltaic effect

When light whose photon energy is larger than the energy gap of silicon falls on silicon, pairs of holes and electrons are created. These current carriers diffuse together through the silicon for a finite time characteristic of the semiconductor. If, however, before recombination they encounter a p–n junction, the field in the junction will be such as to separate them. For example, in Figure 13.1a hole–electron pair is created by absorption in p-type material

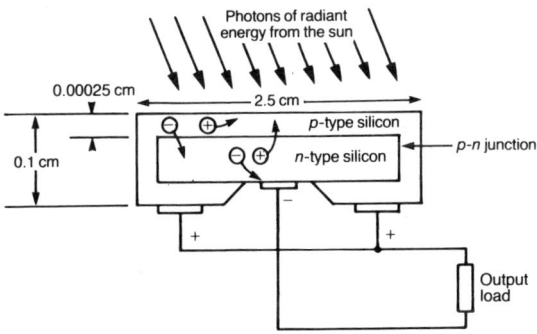

Figure 13.1. Cross section of a photovoltaic (solar) cell. The basic material is single-crystal n-type silicon, with a thin layer of p-type silicon located just beneath the surface. Enough energy is transferred from incoming solar rays to the holes and electrons in the silicon to overcome the junction barrier voltage and to establish current flow in the external circuit

where holes are in the majority. The electron will diffuse into the n-type material and will be accompanied by a flow of charge in the external circuit so that a negative charge appears at the terminal connected to the n-type material. The hole is the minority carrier in the n-type material and on encountering the field at the junction will diffuse into the p-type region so giving a positive charge at the terminals connected to the p-type material. hence when a load is connected at the terminals a current flow is established in the external circuit.

The silicon photovoltaic cell is a useful general purpose detector for use with light sources in the visible and near infrared (0.5-1.0 μm) wavelength range. Filament lamps, sunlight, gallium arsenide and gallium phosphide lamps are all suitable sources.

The cells can be divided into two classes. In one class, an object interrupts a lght beam between a source and the photocell. This class includes punched card and paper-tape readers and counters for objects on assambly lines. In the other class, the cell measures some characteristic of the light source itself — its position in devices for guidance, its intensity in light meters or its modulation to receive speech, television or coded data.

Solar cell

The photovoltaic effect is used in a device known as the solar cell to convert the radiation from the sun directly into electrical energy. The cell has achieved much publicity because of its use as batteries in space vehicles and space satellites. Its advantages in this connection are high power output to weight ratio (+2 W per lb), high conversion efficiency of solar radiation to lectricity (18%), simplicity, ease of fabrication and unlimited life. These properties also make such cells attractive for non-space terrestrial generation at modest power levels especially in remote locations where generation of electrical energy by conventional means is either unavailable or too costly. For instance, many desert regions remain barren because there is no cheap source of energy available for pumping existing underground water to the surface. The abundance of solar energy for which the regions are noted would make this an obvious choice. A solar cell yields about 0.5 V and a current of approximately 0.05 A and hence groups of cells are connected in series to augment the available voltages and groups of series circuits are connected in parallel to increase the current.

An efficiency of 18% represents about half the efficiency of conventional generating plants. In spite of its advantage the solar cell has not moved into serious contention as a source of large amounts of electric power chiefly because of its high cost, the necessity of large collecting land areas and poor duty cycle when terrestrially based. The high cost is due to the low manufacturing yields of high efficiency units and to the high cost of the single crystal material from which the device is constructed.

Since the production of single crystals of the requisite purity is an inherently expensive process this is where gains are due to be made. Thin film technology appears to be a possible answer, for thin films can be produced by mass production techniques and hence significant cost reductions could be brought about. It is of interest to note that to overcome the problem resulting from the poor duty cycle it has been proposed that a large array of solar photocoltaic cells be put into space in equatorial orbit when the sun would shine upon them nearly 100% of the time. The dc power obtained from the cells would then be converted into microwave power, beamed to the surface of the earth and then converted back to d.c. power. The concept, which has become known in the USA as the Satellite Solar Power Station (SSPS), has led to a series of studies of the technology and associated economics and the conclusion is that although technically feasible it is too high in cost to be competitive with established methods of power generation but that it is an option that should be kept open in the event of a cost breakthrough occurring.

Photoemission

The process by which electrons are freed from the crystal lattice and emerge into the surrounding space is known as electron emission. When the energy required to accomplish this is supplied by the use of radiant energy the process is known as photoelectric emission.

Photoemission from a solid occurs when valence electrons are given sufficient momentum by the absorption of radiant energy photons to escape from the solid and hence become a free current carrier. The current resulting from this photoelectric effect is directly related to the number of light photons striking the cathode surface and the ease with which the electrons are ejected from the surface of the semiconductors is related to a surface property of the material known as its work function (ϕ) Electrons will only be emitted if there is sufficient energy in the incident photons to overcome this work function. Hence in order to escape from the surface an electron must have an energy at least $e\phi$ joules. The photon energy is expressed by $h\nu$ where h is Planck's constant (6.624×10^{-34} N m s), and ν the frequency of the light. Hence for emission

$$hv > e\phi$$

Thus it is the frequency of the incident light and not the intensity or brightness which determines whether or not emission takes place.

By the introduction of argon into a photoemissive device an increase in current output can be achieved. The electrons emitted by the cathode in moving to the anode collide with the gas atoms and produce ionisation. The electrons thus freed frofm the atoms join in the charge attracted to the anode. The current may be increased sevenfold over that in an equivalent vacuum cell.

The photoemissionc ell consists of a glass bulb which may be either evacuated or filled with an inert gas, enclosing a photosensitive cathode and an anode to collect the emitted electrons. The cathode is of large area and is so positioned that the light can fall on it. The anode is of small dimension usually a thin rod or wire or metal plate. The photocathode material usually consists of a thin film of alkaline metal such as caesium deposited on a base of silver or silver-plated metal (silver being the best metal conductor known). Frequency of radiation necessary to overcome the work function of caesium is 1.8 eV which corresponds to a wavelength of 436 THz which is very near the lowest possible frequency for light. Hence it follows that light of any frequency possesses enough energy to stimulate emission of electrons from caesium which is the reason for its use in cathode surfaces in emission type potocells.

In caesium the electrons are very loosely held and hence under the influence of radiant energy are readily displaced from the cathodic material and attracted to the anode, so constituting an electric current.

Another composition with high photoemission properties is made from an antimony–caesium compound.

The usual circuitry associated with a phototube is shown in Figure 13.2. Light striking the photocathodic surface transfers energy to the electrons causing them to be ejected and attracted towards the anode which is held at a voltage that is positive with respect to the cathode. current is thus generated while radiation is incident on the cathode.

Since the phototube is a current source that is dependent on the anode–cathode voltage, it is important that the voltage applied

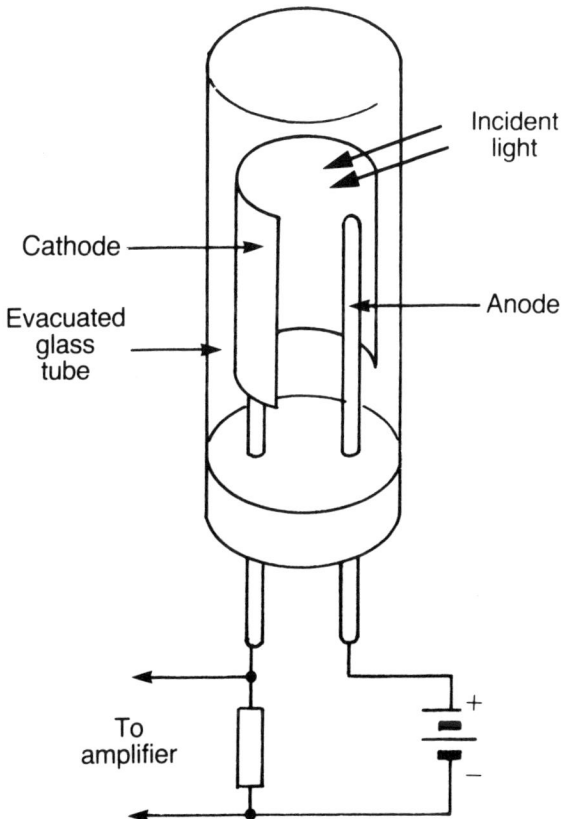

Figure 13.2. Photoemissive tube

between the anode and cathode be as small as possible. If, however, the voltage is too small, the electrons emitted from the cathode will form a cloud or space charge in the vicinity of the cathode and by opposing the field due to the anode potential, will repel the emitted electrons back to the cathode, so prohibiting any flow of current.

In the case of a vacuum phototube if the anode potential exceeds a certain critical value all the emitted photoelectrons will be collected by the anode, any further increase of anode potential will not result in further increase of the photocurrent. This condition is known as saturation and the corresponding value of the photocurrent is termed the saturation current. Photocells are usually designed so that the saturation voltage is as small as possible, the preferred voltage being of the order of 100 V.

Photomultiplier

In addition to the vacuum and gas filled emission cells there is a third type known as the photomultiplier. In this device, amplification of the current is carried out within a high vacuum tube by means of a succession of auxiliary electrodes (Figure 13.3) termed dynodes,

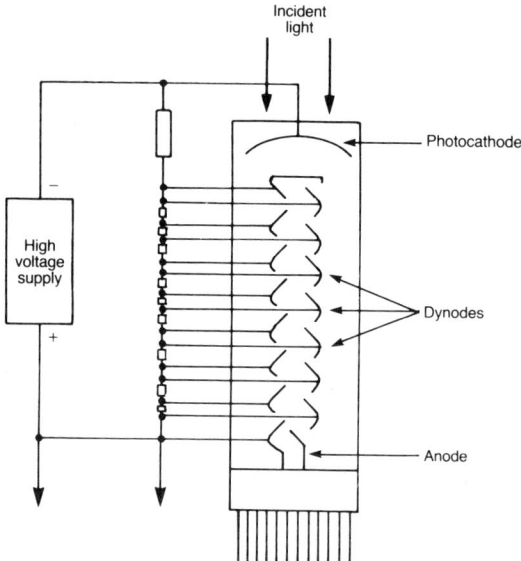

Figure 13.3. Photomultiplier tube

each successive dynode being maintained at a higher potential than the previous one. Electrons emitted by the photocathode are directed to the first dynode by a voltage difference, several secondary electrons being ejected from the surface of this dynode and are accelerated towards the next dynode from which are ejected still more electrons, thus amplifying the current. This process of multiplication continues through each of the dynode stages, until finally they are collected at the anode.

Photomultiplier tubes generally contain about eleven dynodes the electron amplification being of the order of 10^6 to 10^7. The supply voltages necessary to achieve this vary from 1000 to 2000 V. The device is used almost exclusively in photometric work which is the measurement of the luminous intensities of light sources.

Photoconductivity

The phenomenon of photoconductivity deals with the increase in electrical conductivity of semiconductors under the influence of light, in other words their resistance decreases when light shines on them, the direct effect being an increase in the number of mobil charge carriers in the material.

Resistance in the absence of light, known as 'dark resistance', can be very high, in excess of 1 MΩ whereas in the presence of light it can be as low as 100 $\varLambda$. This large difference in resistance caused by light change, with its attendant difference in current flow markes this property of value in switches for the opening and closing of an electric circuit.

The bulk of the semiconductor materials come from groups II and VI, the commonly used compounds being the sulphides and selenides of cadmium and lead. The material used determines the range of frequencies to which the cell is sensitive. Cadmium sulphide is mainly suitable for visible light whereas lead sulphide operates in the infrared region and hence is used for infrared detectors.

Application

The photoconductive effect finds practical application in television cameras, infrared detectors and light meters. Television cameras provide conversion of an illuminated scene into electrical impulses for transmission by means of a photocell. In the Vidicon television camera use is made of cadmium sulphide as the photoconductive cell material, whereas the other type of camera — the Orthicon — uses photoemission based on caesium.

Photodiodes and phototransistors

When a diode is operated in the absence of light, a small reverse current is generated, due mainly to thermal generation of minority carriers near the junction, the forward current remaining unaffected. Radiant energy has the same effect and hence when a diode is exposed to light there is an increase in reverse current. The gain is, however, fractional and amplification is necessary if greater power is required, amplification being effected by a transistor. The response time is, however, very fast, a photodiode switching its current on in a matter of nanoseconds, and hence the device is useful in high-speed counting applications. Another advantage is that they can be made very small, about the size of a pinhead.

The photodiode is provided with a glass window and when light is focused on this window it will respond to the light by generating electronhole pairs. The number of pairs generated is proportional to the intensity of the light and the time exposure. Figure 13.4 shows a simple phototransistor circuit, arrows representing light.

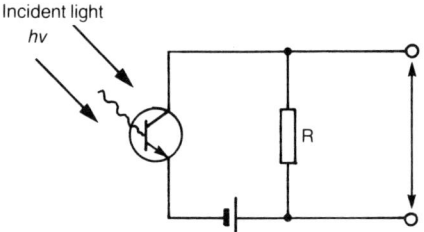

Figure 13.4. Phototransistor amplifying circuit. Signal current induced by incident light is amplified by transistor action

A phototransistor behaves like a reverse-biased photodiode except that as two junctions are present current amplification takes place by normal transistor action. The construction of a phototransistor is similar to that of a normal transistor except that it is so designed that light is cncentrated on the base–collector junction which is reverse biased, the emitter–base junction being forward biased. Light is, however, the controlling input quality rather than the base current. Typically it has a current gain of 100–500 compared with the diode but it has a much slower response time.

Photodetector arrays

Because of the facility with which photodetectors combine with circuitry, complex arrays, both photodiodes and phototransistors are used in immense numbers in industry. These take the form of a two-dimensional array numbering up to several hundred on a single silicon chip. Such configurations are used in image detectors, sensors, optical character recognition (OCR), charge-coupled devices etc.

Electroluminescence

This is the reverse of the photovoltaic effect, optical radiation being the result of an input of electrical energy. The electrical input generates a non-equilibrium excess of charge carriers whose recombination is accompanied by the emission of radiation. This aspect of optoelectronics is of great industrial significance for it forms the basis of what are termed display devices.

Display devices

Electronic display devices are mainly concerned with the visual presentation of small numeric and alphanumeric output information in such mass consumer equipment as desk-top and hand-held calculators, electronic watches and clocks, motor car digital-readout dashboards, cash registers, electronic grocery scales, push-button telephones, cameras, etc. To meet these demands a variety of electroluminescent systems have been introduced, though in recent years established products such as gas discharge devices, incandescent and fluorescent displays have tended to give place to other forms, exploiting the characteristics of more modern sources such as light emitting diodes (LEDs) and plasma display panels. However, these in turn are meeting mounting competiton from liquid crystal displays. Although little more than a laboratory curiosity a few years ago, liquid crystals now have so much potential for low manufacturing and power cost, that they are capturing an ever increasing share of the market. Other potential forms such as light emitting film (LEF), electrophorectic image displays (EPID) and ferroelectric ceramics are other very recent developments.

Luminescence

This can be defined as the emission of optical radiation in the visible range of the spectrum and arises as a result of energy release during electronic transitions within a material. Not all electrons, however, give up their energy as light, for within a material there are regions known as trapping centres which absorb electrons without any emission, the energy being transferred to the crystal lattice in the form of heat. Whether or not energy is released as light is largely related to the composition of the material. Light emitting diodes yield luminescence when voltage is impressed, the light energy source being the *p–n* junction, current flowing across this resulting in photon (light) emission. The amount of light emitted is proportional to the amount of current passing. With a fraction of a milliampere, sufficient light is emitted to be discernible in the dark, 5–10 μA produces enough light to be visible in a shaded area, while 25 μA produced sufficient light to be seen in direct sunlight. Maximum current that may be passed depends primarily on the rate at which heat may be abstracted from the junction region and the most powerful light sources are usually designed with an efficient heat sink in the vicinity of the junction.

Efficiency and brightness

The performance of a LED is measured by the intensity of the emitted radiation and also by the relative response of the eye. The first known as the quantum or conversion efficiency is measured quantitatively by the ratio of the number of photons produced to the number of electrons passing through the diode. A second performance factor, the visual response to the radiation emitted from the diode surface, is a measure of the brightness or luminance and is proportional to the external quantum efficiency of the diode and to the sensitivity of the eye. A high external efficiency requires that the greatest number of photons be incident normal to the emitting surface, whereas high brightness requires all the emission to cover as small an area as possible. Hence these two requirements conflict one with the other. External efficiency and brightness are determined by the internal generation of light at the junction and also by the extraction of light from the crystal which is independent of the light generation processes at the *p–n* junction. For each of these processes there are several loss mechanisms that limit the overall performance. The internal generation of light is highly dependent on the purity of the material, especially in the vicinity of the *p–n* junction where radiative recombination occurs. Defects and contaminants give rise to recombination centres with non-radiative recombination. The trapping centres absorb electrons without any emission due to the recombination energy being transferred to the crystal lattice in the form of heat.

Extraction of the emitted radiation from the crystal is also a major problem. Since the refraction index for a *III–V* group semiconductor is about 3.5, the internal critical angle for light incident at the crystal–air boundary is only 16°. All rays of light striking the surface at an angle exceeding 16° suffer total internal reflection, and as a result, most of the emitted light is reflected back at the crystal. Hence to improve the external efficiency, losses caused by bulk absorption have to be minimized and surface transmission has to be increased. One method of achieving this is to give the crystal a dome structure (Figure 13.5), but this is a very expensive method, for in addition to machining the semiconductor it uses a significant volume of the material.

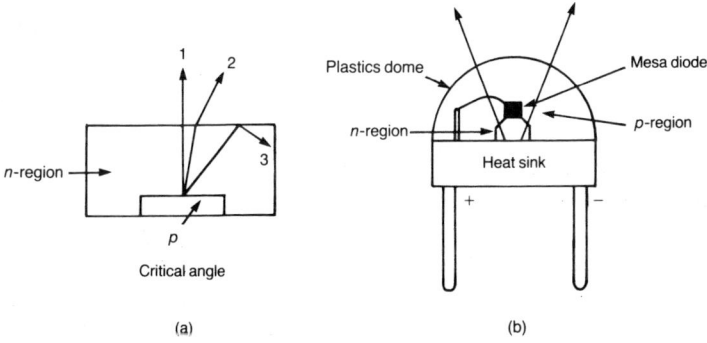

Figure 13.5. In the flat type of diode, light generated at the junction is emitted as visible light only if it follows paths 1 and 2 which are within the critical angle. If light falls outside this angle as shown in path 3, it is reflected back into the diode and is absorbed. To improve the amount of emitted light an arrangement shown in (b) is used. A mesa type diode construction is used, the hemispherical plastics dome effecting a notable improvement in the amount of light leaving the crystal

Hemispherical domes and lenses cast from plastics are effective in increasing the external efficiency by a factor of 2 to 3. In some cases the *n*-type doesn't absorb light originating at the *p*-side of the junction as much as the *p*-type and hence if *n*-type material is used for the dome structure, external efficiency is improved by reduction of the absorption losses.

Materials

As the response of the human eye is limited to a wave length range of between 400 nm (corresponding to a band gap of 3 eV) and 720 nm (1.72 eV band gap), it is apparent that semiconductors with energy band gaps outside these limits are not capable of providing visible emission. It is, however, possible to overcome this handicap by forming alloys between a compound with a small energy gap and a second compound with a large energy cap. LED material is contained within groups *III–V* of the periodic table (see Table 13.1),

Table 13.1 Diode crystal properties

Materials	Colour	Energy (band) gap (eV)	Wavelength (nm)	Luminous efficiency (lm W^{-1})
GaAsP	Red	1.7–2.25	660	42
GaPZnO	Red	2.26	690	20
GaPN	Green	2.3	560	677
GaN	Blue	3.5	570	648
InGaP	Amber	1.34–2.26	617	284
AlGaAs	Red	2.3	675	16

the most widely used being gallium arsenide phosphide (GaAs$_{1-x}$P$_{xx}$) the colour of the light emitted being red. Newer materials such as GaP for red and green emission, IN$_{1-x}$GA$_x$P for yellow and GaN for blue are also becoming available.

LEDs are fabricated from material that is epitaxially deposited on single crystal GaAs or GaP substrates both being available with ingot cross section of up to 3 cm. Both liquid-phase epitaxy and vapour-phase epitaxy are used for preparing LEDs, the former being used for GaP and the latter for GaAsP.

Gallium phosphide

The band gap radiation (Figure 13.6) of GaP which reaches a peak at about 560 nm is very close to the wavelength of maximum eye response. This makes GaP potentially one of the most useful of all visible semiconductor light sources since in addition to green light,

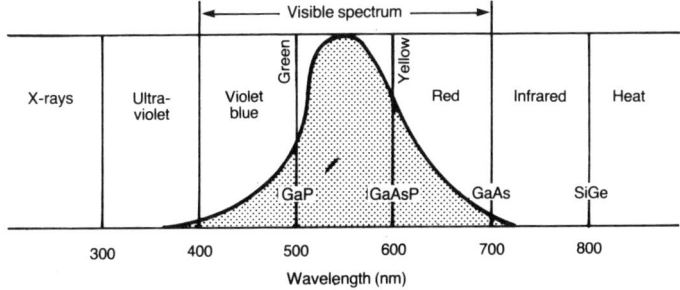

Figure 13.6. A gallium arsenide phosphide junction radiates energy between 600 and 700 nm which is within the visible red, whereas gallium phosphide radiates light energy in the green region. GaAs radiates in the invisible 700 to 800 nm range. The ratio of arsenic to phosphorus largely determines whether the light emitted will be red, green or yellow. As the light efficiency of the junction in the green and yellow regions is low, most LEDs at present aim at a red emission in the region of 650 nm

both red and amber can be produced by appropriate dopants. By modification in the fabrication process the use of these colours side by side can be made possible in display devices. GaP is prepared by allowing gallium to react with phosphine which can be prepared in a high state of purity by decomposition of aluminium phosphide with water. This results in gaseous phosphoretted hydrogen whcih after being dried and cleaned is passed over gallium in a furnace at 1200 °C. In this state the GaP crystals exhibit only weak luminescence and to produce a product which will luminesce satisfactorily, the crystals have to be doped with either zinc or zinc plus oxygen. The former gives rise to the emission of green light while zinc plus oxygen gives rise to red light, and it is the latter that is mainly of interest.

For the manufacture of a satisfactory LED product, it is necessary to convert the polycrystalline material into a single crystal. In the case of GaP diodes, this is accomplished by the process of liquid phase epitaxy (the term being derived from a combination of the Greek words 'epi' meaning upon and 'taxos' meaning arranged) and relates to the deposition of a solid layer of the material on an underlying substrata. Presence of nitrogen is necessary to obtain room temperature green emission. Red emitting diodes of GaP are commercially availble with external quantum efficiency as high as

3%. It is also possible to obtain orange or yellow emisison by the introduction of N and O into the n- and $p$sides of the junction respectively. Since green emission arises from the n-side and red emission from the p-side of the junction the same diode can simultaneously emit both colours whose translation by the eye yields an orange or yellow appearance. A new technique for forming GaP crystals is the synthesis solute diffusion (SSD) process whereby crystals of GaP are grown under phosophorous vapour pressure of about one atmosphere and at relatively low temperatures. Crystals as large as 40 mm in diameter and weighing 150 g can be grown.

Gallium arsenide phosphide
It may be said that at present numeric and alphanumeric displays are dominated by $GaAs_{1-x}P_x$ light emitting diodes. The lead that this material now enjoys in the major application for LEDs—seven segment numeric displays* is due primarily to its lower cost and also to the fact that it is brighter to the eye. It has also many other desirable features viz, reliability, long life, compatibility with ICs, fast response time, operation over a wide range of temperature and high contrast ratios.

GaAsP is manufactured (Figure 13.7) by the interaction of gallium with a hydrochloric acid/hydrogen mixture at 850 °C. The resulting gallium monochloride flows out of the reaction chamber and mixes with a Ph_3/AsH_3 dopant and then passes over a GaAs substrate where it reacts with its surface to form a single crystal epitaxial deposit of GaAsP. The donor impurity is usually Se which is introduced as H_2Se. For use the GaAsP slice is scribed and cleaved to give dice about 600 μm square the actual displays being made up from individual dice.

Multiplexing systems in general use, comprise a complete display of digits driven from a single decoder-driver instead of each character having its own decoder driver. The single decoder-driver applies periodic current pulses to each character in turn in the complete display. In general, it is economical at present to multiplex a display with eight or more digits, but with the cost of multiplexing

---

*In the type of display most commonly used numbers are constructed from seven bars or segments similar to a clock number so that by illuminating the proper combination of segments all the numbers from 0 to 9 can be formed.
A selection from the alphabet can also be formed.

Electroluminescence 243

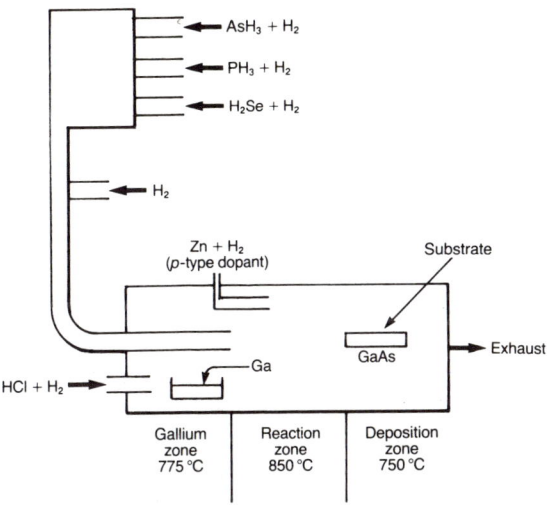

Figure 13.7. Vapour-phase gallium as the monochloride is obtained by passing hydrogen chloride over the molten metal. Arsenic and phosphorus are formed by the thermal decomposition of arsine and phosphine, respectively, hydrogen being used as the carrier gas. As the gases pass over the substrate they react with its surface to form a crystal epitaxial deposit of GaAsP

falling the probability is that it will soon be economical to strobe as few as four or five digits. In addition to reducing the number of leads needed to connect the display, a pulsed LED appears brighter to the eye than a continuously driven one.

Liquid Crystals
These materials have unique optical and physical characteristics and have so much potential for low cost manufacturing as well as minimal power consumption that they present a severe challenge to LEDs.

The compounds used for LCDs are a class of organic chemicals known as nematic, and differ from ordinary crystals in that they possess the unusual property of changing from a normal transparent to an opaque state when exposed to an electric current (Figure 13.8). Thus, information can be displayed by selectively causing this change to occur in certain areas of a device in a similar manner to the seven-segment arrangement used in LEDs. It is to be noted that light

244 Optoelectronic components

is not generated but modified in such a manner as to give rise to a visual display.

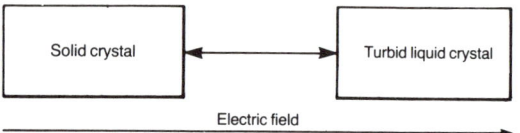

Figure 13.8. Liquid crystal behaviour exists when an electric field is applied, turbulence being generated causing the transparent material to become opaque. The effect lasts only as long as the field is applied

Displays are made by putting a thin layer of the liquid crystal (Figure 13.9) between two plates of glass, each of which is coated on the inside with a conductive coating such as tin oxide. One plate is etched with the segments required to produce the necessary lettering or numbers, an electric voltage being applied selectively to the

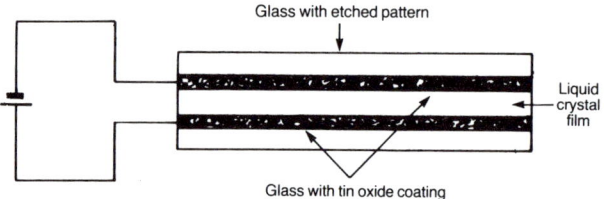

Figure 13.9. The device consists of two parallel glass plates with a film of the liquid crystal sandwiched between them. A conductive coating such as tin oxide on the inside surface of the plates, each with a lead, completes the fabrication

segments needed to produce the desired characters. The opaqueness is not a chemical reaction, the turbulence and light scattering effect resulting from the application of the electric field. When the field is removed the liquid crystal material returns to its transparent condition.

Liquid crystal material is inexpensive and because it is used in very thin films cost are very low. Costs as low as 30p per digit have been estimated. Since the display operates by reflecting and not generating light, power costs are also very low, typically measured in micro-watts per square centimetre. However, there are limitations. One is that temperatures must be kept above 0 °C and below 60 °C.

Abother is the comparatively long response time, that is the period which it takes for the display to revert back to its initial state after the field is removed. This varies between 50 ms and 1 s depending on the temperature and the material used. Multiplexing is also a problem because LCDs do not have a sharp threshold (pulse height selection). Some ingenious solutions to this problem have been proposed — incorporation of MOS counter decoder-driver chip into the display itself and the application of a circuit in which multiplexing is done externally.

The advantage of multiplexing is illustrated in the case of an experimental eight-digit liquid crystal display. A conventional display without multiplex requires some 65 electronic drivers and leads. The multiplex version requires only 16 electronic drivers and leads.

Coloured displays

It is possible by the incorporation of certain dyestuffs to produce coloured displays. This depends on the ability of the liquid crystal to orient the dye molecules whose absorption depends on direction, a change of light absorption being made possible by the application of the electric field.

Table 13.2 Operational characteristics of display devices

Characteristics	LED (GaAsP)	Liquid crystal	Plasma	Gas discharge tube
Voltage	1.5–5	15–50	150–200	150
Current	75–150 μA	10–20 μA	25 μA	1–5 μA
Cost per digit (very approx)	50p	30p	80p	85p
Temperature range	–50 °C to +100 °C	10–55 °C	—	—
Digit height (mm)	3–6	13	12	12

Plasma displays

It would seem that a dozen or so digits are about the upper limit for LEDs. Above this figure plasma (ionised gases) displays are economically attractive. The characteristic neon organge red colour of plasma displays gives a pleasing readout and is easy on the eyes.

The chief drawback is the high voltage required to operate it, in the region of 170 V. Plasma displays are fabricated 'sandwich fashion' from a rear cover containing thick film electrodes which act as cathodes and thick film dielectrics, a spacer frame and a front glass cover containing transparent anodes.

Fibre-optics

Fibre-optic communications by means of which information is transported in the form of light signals conducted along a glass film (Figure 13.10) may be listed as an important advance of the 1970s.

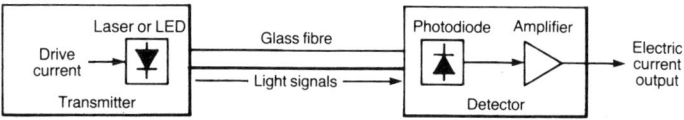

Figure 13.10. Basic fibre transmission system

The optical fibre is a versatile transmission medium and may be used in a variety of applications where twisted copper wire pairs, coaxial cables and metallic wire guides are now used for transmission of information. Its application ranges from short data links within a building to long telecommunication trunk circuits connecting offices within a city and between cities. The small size of the individual fibre (100 μm) the large information capacity, freedom from electromagnetic interference and the potential economy are some of the features which made optical-fibre systems more attractive than copper wire systems. The system is of interest not only to telecommunications but also for application in aircraft and instrumentation.

Basic fibre-optic system

The glass fibre consists of a core contained in a cladding of lower refractive index than the core. Light conduction in the core results from internal reflection at the interface between the core and the cladding (Figure 13.1). For efficient internal reflection the angle of incidence at the interface must be smaller than the critical angle

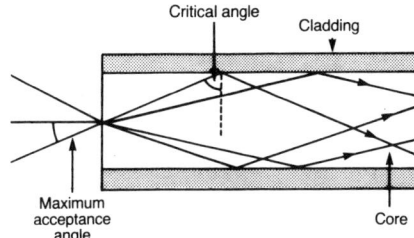

Figure 13.11. Wave guidance of incident ray by internal reflection in a glass fibre

which is determined by the refractive indices of the core and the cladding. If the refractive indices differ by 1% the critical angle is about 8°. If the incident ray is greater than the critical angle it will not be internally reflected but will be refracted into the cladding. The three basic requirements of a fibre-optic system are:

(1) A transducer that converts the electrical input signals into light for conduction into the glass fibre.

(2) Optical fibre light guide

(3) A detector to convert the received light back into electrical signals.

(1) Light source

At present injection lasers and LEDs are the major light sources used in fibre-optic systems. Both sources radiate in the near infrared (0.8–0.9 μm) and can be directly modulated by varying the injection current so that the information to be conveyed can be incorporated directly into a light signal. The diode material can be doped appropriately to ensure the wavelength of the emitted radiation is located at or near a minimum in the attenuation of the glass fibre. In the case of the LED the recombination radiation is emitted incoherently from the junction, whereas in the case of the laser the coherent radiation produces a narrower emission angle resulting in a more efficient coupling of light power to the fibre. The laser most frequently used is the GaAlAs laser which consists of a number of epitaxially grown layers of the mixed-crystal AlGaAs on a GaAs substrate. In a material of this type the input current injects carriers across the p–n junction which converts the electrical signals into a steam of light pulses. Only a very low drive current (100–200 mA) is required. A GaAs laser generates a beam of light only 2 μm in diameter which is small enough to couple all the laser power into the thinnest of single mode optical fibres.

(2) The Optical Fibre

The wave propagation characteristics of the fibres are determined by the size of the core, its refractive index distribution and the refractive index of the cladding. Essentially, all the guided energy is required to be confined to the fibre core and hence for efficient light transmission as little radiation as possible should be refracted into the cladding, loss due to absorption and scattering being kept to a minimum. This requirement is met by using a core of higher refractive index than the cladding which gives internal reflection at a low angle of incidence.

In fibre technology one critical parameter is the fibres' attenuation measured in the decibel loss per km of length. To reduce the attenuation to a minimum necessitates the use of repeaters or amplifiers which have to be included in the cable. Loss in modern equipment has been reduced to less than 2 dB km^{-1} at a wavelength of 0.9 μm and less than 1 dB km^{-1} at 1.05 μm, that is one-third of the light is lost in passing through 1 km of fibre. Essential to the achievement of low loss has been the elimination of both scattering and absorption of light energy as it travels through the fibre. Causes of scattering are non-uniformity of the fibre and particles and bubbles trapped in the film during fabrication.

Fabrication of fibres

The central region, the core, typically consists of doped silica glass produced by a method of chemical vapour deposition. Silicon tetrachloride gas is reacted with oxygen at a high temperature (1000 °C) to form silica glass which is deposited on the inside surface of a tube of pure silica glass. This tube eventually becomes the cladding of the glass fibre. Typically it might be 2 cm in diameter and 50 cm long and is known as a preform.

Included in the gas stream are small amounts of such dopants as germanium or phosphorus to raise the refractive index by about 1%. After a sufficient amount of material has been deposited, the preform is heated to a higher temperature (2000 °C) at which it softens and eventually collapses to form a rod. This is fed into an electric furnace and a fibre is drawn at a rate of the order of 1 m s^{-1}. Several kilometres of fibre can be drawn from the 2 × 50 cm preform. A protective coating of plastics material to protect the surface from mechanical abrasion and atmospheric attack is then applied. There are four types of fibre:

(a) Single-mode fibre. This is composed of a low-loss fibre with a very small core (2–8 μm). It necessitates a laser source for the input signals because of this very small entrance aperture. The small core radius approaches the wavelength of the source and hence only a single mode is propagated (see Figure 13.12a).

(b) Step-index multimode fibre. This type of fibre has a much larger core (25–75 μm) than the single-mode core and permits non-axial rays or nodes to propagate through the core compared with only one mode through a single mode fibre (see Figure 13.12b).

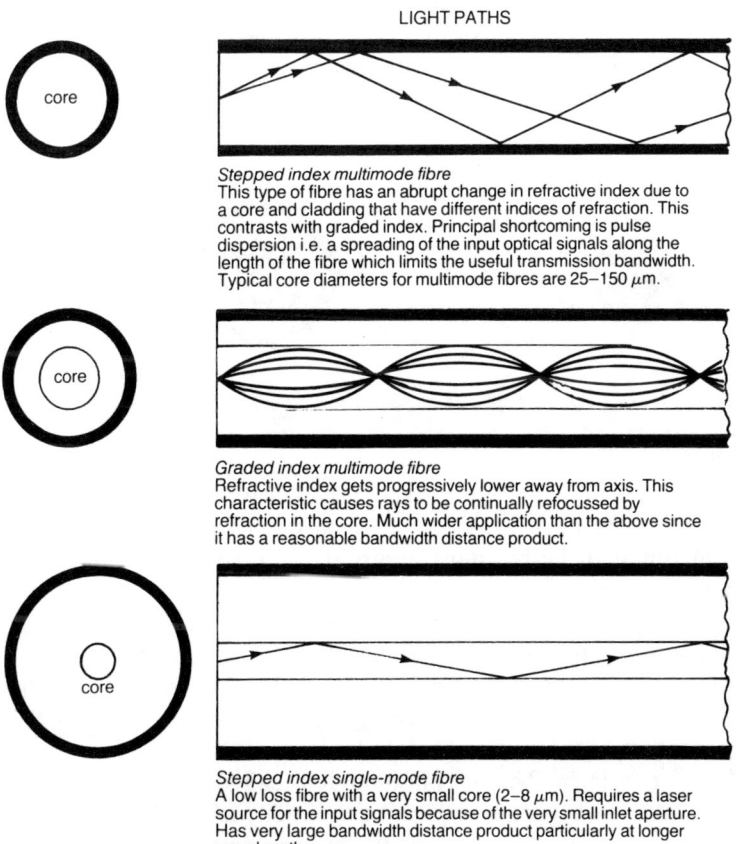

LIGHT PATHS

Stepped index multimode fibre
This type of fibre has an abrupt change in refractive index due to a core and cladding that have different indices of refraction. This contrasts with graded index. Principal shortcoming is pulse dispersion i.e. a spreading of the input optical signals along the length of the fibre which limits the useful transmission bandwidth. Typical core diameters for multimode fibres are 25–150 μm.

Graded index multimode fibre
Refractive index gets progressively lower away from axis. This characteristic causes rays to be continually refocussed by refraction in the core. Much wider application than the above since it has a reasonable bandwidth distance product.

Stepped index single-mode fibre
A low loss fibre with a very small core (2–8 μm). Requires a laser source for the input signals because of the very small inlet aperture. Has very large bandwidth distance product particularly at longer wavelengths.

Figure 13.12. The three types of optical fibre

250 Optoelectronic components

(c) Graded-index fibre. This is an optical fibre with a refractive index that gets progressively lower away from the axis (Figure 13.12c). This characteristic causes the light rays to be continually refocused by refraction in the core. This is low-loss multimode fibres with a graded index profile would seem to be the dominant type for use in telecommunication systems.

Cables

Fibres have to be formed into cables to protect them during installation. Some thirty or forty fibres combine to give a cable of very large transmission capacity. In addition to external protection, materials have to be incorporated to provide tensile strength. Such strength members are usually composed of steel strands or aluminium (Figure 3.13).

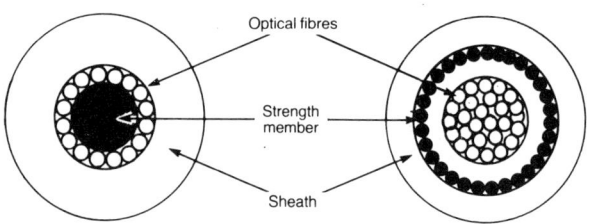

Figure 13.13. Structure of optical fibre cables

(3) Detector

At the receiving end of the system the light waves have to be detected and converted back into electrical signals. Silicon photodiodes have been found to be the most appropriate detectors for fibre-optic systems.

A photodiode is a *p–n* junction which is reverse biased. The junction is exposed to the light ray through a 'window' which is transparent to the radiation and allows the light to penetrate the diode. The quantum efficiency of the diode, which indicates the proportion of incoming photoelectrons converted to current, is in the oder of 90%. The *pin* photodiode is suited for short distance applications where receiver sensitivity is not critical. The avalanche photodiode increases sensitivity by multiplying the primary photocurrent internally resulting in a larger output current.

In both types of photodetector the quantum efficiency is determined by the absorption coefficient of a depletion region with a high electrical field between the two semiconductor regions. The absorbed photons result in the generation of electron–hole pairs. In the depletion region the electric field separates the electron–hole pairs (electrons go to the n-region, holes to the p-region) and this has the result that a current, the photocurrent, starts to flow in the circuit in which the diode is incorporated. The device is usually operated with reverse bias of a few tens of volts. In most cases the current emitted by the detector is very weak and hence has to be amplified.

Non-avalanching Photodiodes

Front illuminated $p + I + n$ structures are commonly used in silicon photodiodes and provide a fast response time and high quantum efficiency.

pin diodes (Figure 13.14) consists of an $n+$ substrate, an intrinsic (i) depletion region and a thin $p+$ region. For applications in the wavelength range 0.75–0.9 μm the required depletion layer width

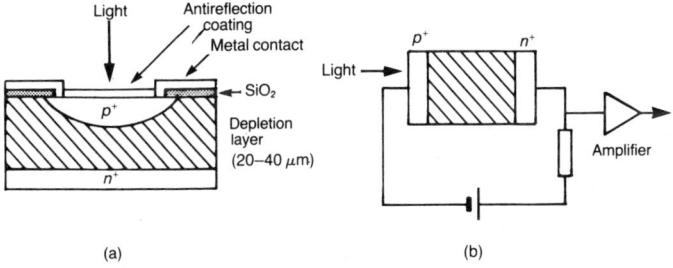

Figure 13.14. (a) Cross section of a pin photodiode. The surface is given a protective layer of anti-reflection material to minimise reflection of the radiation to be detected. (b) Simplified circuit for a pin photodiode

would be about 20–40 μm. There is no internal amplification of the photocurrent, one electron being generated per incident photon or about 0.5 AW^{-1} of optical power emitted by the detector. *pin* photodiodes are biased at 16–20 V.

Avalanche Photodiodes

Avalanche photodiodes (APDs) combine the detection of optical signals with amplification of the photocurrent. The doping profile

(Figure 13.15) is adjusted to result in a p+ substrate, a region with a very high electrical field and a narrow p–$n+$ -region. Internal gain is realised through avalanche multiplication of carriers in the high-field region of a highly reverse biased junction where the photocarriers are accelerated to velocities which are sufficiently high

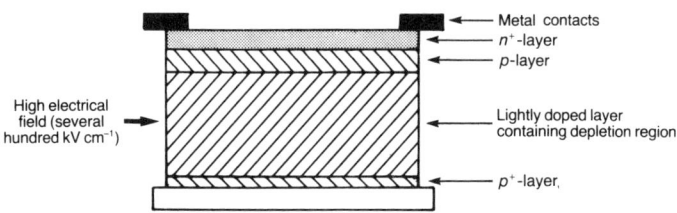

Figure 13.15. Cross section through an n^+ pip^+ avalanche photodiode

to generate new electron–hole pairs through the process of collision ionisation. These new carriers can in turn generate additional pairs. Avalanche diodes operate with a bias voltage between 100 and 400 V and a photocurrent gain of ± 100. The current gain of the APD fluctuates due to the fundamental statistical nature of the avalanche multiplying process and gives rise to noise which tends to degrade the signal.

Recently a light detector has been developed to operate a telephone solely from light pulses transmitted through a fibre. The photodetector detects incoming light signals at one frequency and transmits outgoing signals at another frequency allowing two-way transmission through a single fibre.

This means that a single device coupled to the end of a glass fibre can serve as both a transmitter and receiver of light wave signals. It converts 56% of the incoming light to electric energy, and in fact it is quite possible that one day a telephone with no external electrical connector not even for ringing will be developed, the telephone being operated solely with the power of light transmitted over glass fibres.

Applications

The main application of fibre optics is in the telecommunications industry. The reason is that existing telephone cable conduits are fully loaded and cannot carry many more individual wires. Glass fibres can carry many more conversations than a wire because of the very much higher carrier frequency of a light beam compared with an electrical carrier signal along a wire. Further their dimensions are very small, they use little energy and are easy to modulate by the drive current. Transmission of telephone conversations along glass fibres is already in operation in the UK, USA, Japan and elsewhere.

In the USA many short distance links are in operation and in Indiana a 4.5 km link has been installed between two exchanges which is designed to carry 670 simultaneous telephone calls on a single pair of glass strands. In Japan, twelve cities including Tokyo are linked by fibre optical telephone networks and several houses can receive cable television with two-way communication. In the UK several million pounds worth of optical fibres for the public telephone system is already in operation.

Another obvious use is the guiding of light into inaccessible places for inspection by the naked eye. Such light guides can be incorporated in medical endoscopes for internal examination replacing the tiny electric bubble hitherto used.

Light guides are also being used in computers. Light signals can be transmitted faster along light guides than electrical signals along wires because of the absence of electrical impedance. In addition light guides are also being used for card reading in automatic data processing equipment.

It is evident that if optical communications are to become commonplace the system will have to be integrated and produced much in the same way as ICs. A typical integrated optical IC will contain the light source, modulator and detector on thin film chips and will make them the primary candidates for long distance optical telecommunications.

The modulator format for optical fibre systems can be either analogue or digital. The former has the advantage of simplicity and eocnomy but the large signal-to-noise ratio limits its use to relatively low bandwidths and short distance applications. Pulse-code modulation (PCM) offers the most flexible and efficient means of achieving improved noise performance. Consequently, most of the work of fibre-optic telecommunications is based on digital transmission systems.

Further reading

1. *Philips Technical Reviews.* Gloeilampenfabrieken, Eindhoven, Netherlands.
2. *Systems Technology.* Plessey Co. Ltd., Ilford, Essex.
3. *Spectrum.* Published monthly by IEEE, New York.
4. *Microelectronics, Scientific American.* September 1977.
5. *New Scientist.* Published weekly by IPC Magazines, London.
6. *Digital Integrated Circuits, Operational Amplifiers and Optoelectronic Circuit Design:* Prepared by Texas Instruments Inc.
7. R. G. Hibberd, *Solid State Electronics.* Texas Instruments Inc.
8. W. R. Runyan, *Silicon Semiconductor Technology.* Texas Instruments Inc.
9. Hovill, R. L. and Walton, A. K. *Elements of Electronics for Physical Scientists,* Macmillan.

Index

Acceptor atoms, 7
Active components, 12, 14, 61, 83, 99
Admittance, 159
Alloy-junction transistor, 122
ALU, 187
Amplification,
  common-base, 93
  common-collector, 92
  common-emitter, 88
  principle of, 86
Amplifiers,
  bipolar, 83
  efficiency, 87
  long-tailed pair, 147
  transistors, 86
  unipolar, 99
Analogue circuitry, 143
AND gates, 169
Arithmetic, computer, 201
Arrays, 236
Atomic structure, 1
Attenuation, 161, 248
Avalanche diode, 70, 250

Band structure, 2
Band width, 2
Barrier region, 8
Biasing, 8
Binary notation, 164
Binary system, 164
Bipolar transistor, 83
Bistable circuit, 177
Bit, 164, 191

Block diagram, 15
Breakdown voltage, 11, 67
Bridgman-Stockbarger, 121
Byte, 191

Capacitance, 42
Capacitors,
  ceramic, 50
  electrolytic, 46
  paper, 48
  plastic, 49
Ceramics, 28
Characteristic,
  voltage-current, 77
  transistor, 94
Charge carriers, 8, 84
Charge coupled devices, 111
Choke coil, 56
Circuit diagrams, 14
COBOL, 198
Coercivity, 26
Coils,
  inductor, 54
  transformer, 56
Collector, 83
Collector current, 83
Colour codes, 37
Complementary MOS transistors, 108, 173
Common-mode, 148
Computers, 185
Conduction band, 4
Core, 246

255

256 Index

Crystal growth, 120
Crystal structure, 5
Crystals, liquid, 243
Curie point, 42
Current gain, 85
Czochralski, 121

Dark resistance, 235
Darlington pair, 97
Data processing, 185
Decibel, 248
Depletion layer, 8, 103
DIAC, 81
Dielectric, 42
Digital, 164
  logic, 179
Diodes, 9, 61
Display devices, 237
Donor atoms, 6
Doping, 6, 121
Drain electrode, 100
Drift, 148

Eddy currents, 59
Electro-luminescence, 237
Electromagnetic field, 12
Electron, 4
Electron beam, 137
Electronics, basic aspects, 1
Emission, 231
Emitter, 83
Encapsulation, 29
Energy bands, 2
Epitaxy, 138, 241
Extrinsic, semiconductors, 5

Fabrication, 118, 129, 132
Farad, 43
Ferri-magnetism, 26
Ferro-magnetism, 26
Ferrite magnets, 27
Fibre optics, 246
Field effect transistors, 99
Figure of merit, 55
Filters, 158
Flip-flop, 177
FORTRAN, 198
Forward bias, 10
Frequency, 158

Full-wave rectification, 63

Gallium arsenide, 127, 224
Gallium phosphide, 240
Gates, logic functions, 168
  AND, 169
  NAND, 170
  NOR, 171
  NOT, 168
  OR, 169
Germanium, 118
Gunn diode, 68

Hardware, 186
Heat sink, 95
Henry, 52
Hole, 4
Hybrid circuit, 132, 139

Image sensor, 117
IMPATT device, 70
Impedance, 59, 156
Impurity atoms, 5, 121
Impurity levels, 5
Inductance, 51
Inductor, 54
Insulated gate, MOST, 104
Insulators, 27
Integrated circuits, 17, 129
  fabrication, 118
Integrated injection (logic) (I^2L), 175, 210
Intrinsic semiconductors, 5
Ion implantation, 126
Ionization, 126

JFET, 103
Josephson junction, 225
Junction gate FET, 103
Junction transistor, 7, 83

Language, computer, 196, 211
Large scale integration (LSI), 18
Laser, 247
Lattice, crystal, 59
Light emitting diode (LED), 237, 247
Liquid crystal displays, 243

Index 257

Lithography, 124, 135
Load line, 96
Logic, gate circuits, 166
   DTL, 180
   ECL, 182
   MOS, 172
   RTL, 179
   TTL, 181
Long-tailed pair, 147
Luminescence, 237

Machine language, 196
Magnets,
   permanent, 26
   soft, 26
Magnetic bubble memory, 191, 225
Magnetic flux, 53
Magnetic materials, 26
Main frame, 191
Majority carriers, 8
Masking, 124
Memory, 114, 189, 208
Metals, 21
Microelectronics, 19
Microprocessors, 204
Microsensor, 214
Microwave device, 69
Minority carriers, 8
MNOST, 108
Monolithic circuit, 132
MOS-FET, 104
Multiplex, 242

NAND gates, 170
Negative resistance, 74
Noise, 183
n-p-n transistor, 84

Ohm's law, 32
Operational amplifier, 146
Optoelectronics, 115, 228
OR gate, 169
Oscillator, 68

Passive components, 12, 31
Permeability, 26, 53
Permitivity, 46
Photocell, 232

Photoconductivity, 235
Photodiode, 235, 250
Photoemissive, 231
Photolithography, 124, 135
Photo-mask, 124
Photomultiplier cell, 234
Photon, 229, 238
Phototransistor, 235
Photovoltaic effect, 229
Pinch-off voltage, 103
Planar technique, 123
Plasma displays, 245
Plastics, 29
p-n junction, 7
p-n-p transistor, 84
Potentiometer, 39
Power rating, 34
Printed circuit, 16
Program, 187
Pulse code modulation, 145

Q-factor, 55

Random access memory, 221
Read-only memory, 190
Rectification, 62
Rectifier, 62
Registers, 187
Reluctance, 53
Resistance, 31
Resistors, 35
   composition, 35
   metal film, 36
   wire-wound, 36
Resonance, 55
Reverse bias, 9
Ripple, 64

Saturated mode, 179
Schottky, barrier diode, 72
Semiconductor materials, 22, 24
Semiconductor preparation, 118
Semiconductors, 25
Shift resistor, 113
Signal, 14
Silicon, 119, 223
Silicon chip, 126
Silicon wafer, 122
Software, 196, 210

Index

Solar cell, 230
Solenoid, 52
S.O.S. technique, 107
Source electrode, 100
Speed-power factor, 183
Sputtering, 141
Stabilisation, 91
Storage, computer, 189
Super-alpha pair, 97
Superconductivity, 34
Symbols, x
Switching, 73, 167

Temperature coefficient, 33
Thermal runaway, 100
Thermistor, 40
Thick and thin film circuits, 139
Thyristors, 76
Transformer, 56
Transistors,
   action, 84, 89
   amplification, 86
   basic configurations, 87
   bipolar, 83
   field effect, 99
   junction, 7, 83
   MOS, 104
   n-p-n, 84
   p-n-p, 84
   unipolar, 99
TRAPATT mode, 71

TRIAC, 80
Triode (thermionic) valve, 22
Truth tables, 168
Tunnel diode, 73

Unipolar transistor (F.E.T.), 99
Unijunction transistor, 74

Valence band, 3
Varactor diode, 71
Very large scale, integration, 19, 216, 226
Voltage comparators, 153
Voltage divider, 91
Voltage gain, 90
Voltage regulators, 154
Voltage, threshold, 107

Work function, 231

X-ray lithography, 137

Yttrium iron garnet, 192

Zener diode, 67
Zone refining, 119

20. DEC. 1990